AF404726

Thermodynamics, Kinetics and Microstructure Modelling

Thermodynamics, Kinetics and Microstructure Modelling

Simon P A Gill

School of Engineering, University of Leicester, University Road, Leicester, LE1 7RH, United Kingdom

IOP Publishing, Bristol, UK

ISBN   978-0-7503-3147-0 (ebook)
ISBN   978-0-7503-3145-6 (print)
ISBN   978-0-7503-3148-7 (myPrint)
ISBN   978-0-7503-3146-3 (mobi)

DOI   10.1088/978-0-7503-3147-0

Version: 20220301

IOP ebooks

British Library Cataloguing-in-Publication Data: A catalogue record for this book is available from the British Library.

Published by IOP Publishing, wholly owned by The Institute of Physics, London

IOP Publishing, Temple Circus, Temple Way, Bristol, BS1 6HG, UK

US Office: IOP Publishing, Inc., 190 North Independence Mall West, Suite 601, Philadelphia, PA 19106, USA

Cover image: Surface hardening of steels. Image credit: Leica-microsystems.

To my mother.

Contents

Preface

This is a textbook aimed at later stage undergraduates and early stage postgraduates interested in developing a deeper understanding of materials and materials modelling. The contents were originally created as two separate taught modules for first year PhD students, the first on thermodynamics and the second on kinetics and microstructure modelling. This is mirrored in the two part structure of this book. The students I taught were all working on problems in the processing of engineering materials. The courses were initiated so that they should have a fundamental understanding of the theory behind how materials form and evolve, even if their projects were entirely experimental. The two courses, like this book, therefore focussed on engineering alloys. My aim in creating these modules was to make the required concepts as simple and clear as possible, to avoid long, heavily theoretical lectures, and to keep the students active and involved through the use of practical examples. This process was massively helped when I discovered the free demo-version of the PANDAT software, which is perfect for plotting phase diagrams from thermodynamic database (TDB) files, which you can create yourself. This software is therefore a key element of this book. It allows students to generate results themselves and see at first hand the effects of key parameters. The results are illustrated in the book, but it is strongly encouraged that readers use the software to follow the step-by-step instructions to do the examples themselves. The two modules were taught in a computer lab intensively over one week each. The sessions were a combination of short lectures introducing a particular aspect of the theory followed by practical sessions where the students used PANDAT to explore the practical results of the theory. The book follows this structure, starting with the most basic concepts and building on them throughout. As such, the book is intended to be read from start to finish, as opposed to a reference book where you might just look at a single chapter on a particular topic of interest. Additionally, this is not a materials science book, in the sense that little reference is made to specific materials. It is largely presented in generic terms, with illustrations on how to apply the approach to real materials. The book refers to a number of excellent textbooks for readers to obtain further information on specific materials if they wish.

Simon P A Gill
29 October 2021

About the author

Simon P A Gill

 Simon P A Gill is a Professor of Theoretical Mechanics at the University of Leicester in the UK. He is also Director of a Doctoral Training Centre for 65 PhD students on Innovative Metal Processing. He has 30 years experience of numerical and analytical modelling of different materials systems, ranging from ceramics, metals and semiconductors to applications in geology and biology. These studies range across a wide range of length and time scales, from nanometres to kilometres, and from seconds to millions of years. The foundation of many of these studies has been the development of a novel variational principle which allows analytical approximations to complex problems to be generated. This approach is used in this book to derive fundamental expressions for the evolution of microstructural features using basic physical principles, hopefully without them being obscured by the mathematics.

IOP Publishing

Thermodynamics, Kinetics and Microstructure Modelling

Simon P A Gill

Chapter 1

Introduction

The development of new materials has been critical to the progress of human civilization, to the extent that they are even used to define periods of our history, such as the stone age, bronze age, iron age and now the silicon age. This development has been greatly accelerated in recent years by advances in techniques for measurement, processing and prediction. Modelling the evolution of material microstructure is an essential part of this process. Atoms try to organise themselves within a material to minimise their energy, of which thermodynamic energy plays a major role. However the ability to minimise this energy is limited by the kinetic mobility of the atoms to follow this idealistic pathway. Consequently, the first part of this book introduces the concept of thermodynamics in relation to the formation of metal alloys, and the second part builds on this knowledge to couple thermodynamics with material transport kinetics to generate predictive models for the formation and change of material microstructures over time. This introductory chapter briefly introduces the main features that appear in a metal alloy microstructure and their importance in determining the final properties of a structural engineering material.

1.1 Motivation

The importance of developing new materials to support new technologies has been critical throughout human history. In the past this development has often been through a process of trial and error. In the last few decades the role of modelling in the development of materials has become more prevalent. This is largely because the pursuit of materials with improved properties through purely experimental means can be extremely expensive, both in terms of time and money. Physical modelling of material systems provides quantifiable insight into the underlying mechanisms that make them what they are. It also offers a faster route for development, whereby numerical experiments (simulations) can analyse a wide range of options, and hence augment the experimental decision process.

The main objective of materials research is to create an optimal material for a given task, with certain properties, be they structural, electrical, thermal etc. The principle factors that affect the properties of a material are shown in figure 1.1. Here we focus mainly on metals. A metal that contains a combination of two or more elements is known as an **alloy**. The most fundamental inputs that determine the properties of an alloy are:

- the elemental **composition**, e.g. iron (Fe) and carbon (C) are key elements in the formation of the alloy steel.
- the thermomechanical **processing** that it undergoes in its creation, e.g. processes such as casting (as the solid metal forms from a liquid), extrusion or rolling (as the solid is deformed, often at high temperature), and annealing (where the metal is held at a high temperature for a given time, often to relieve some of the effects of the thermomechanical processing).

Together these two factors determine the **microstructure** of the material. The properties of a material, e.g. yield strength, are then entirely dependent on that microstructure. **Microstructure evolution** is the process by which a microstructure changes over time. These changes can occur rapidly during formation or processing (over time scales of seconds to hours) or very slowly when in service (over many years or decades). The two key contributions that determine how a microstructure evolves are:

- **thermodynamics**—the interaction between thermal energy, mechanical energy and chemical energy. The composition of an alloy strongly influences its chemical energy and the resulting crystal structures that form. Thermomechancial processing inputs thermal and mechanical energy to the material.
- **kinetics**—the transport or rearrangement of atoms within a material. Kinetic processes are much more rapid at higher temperature and have a strong

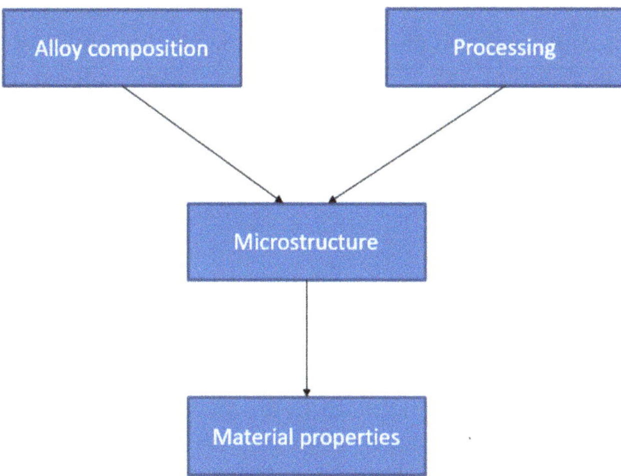

Figure 1.1. The stages of materials modelling.

influence on how quickly changes in a material can happen. Here we focus mainly on diffusive processes. Some elements can diffuse rapidly (e.g. carbon) and some are orders of magnitude slower (e.g. iron).

The rest of this chapter proceeds as follows:
- the main features of alloy microstructures are introduced in section 1.2.
- Section 1.3 relates these features to the resulting structural properties of the material.
- Section 1.4 describes the roles of thermodynamics and kinetics in the formation of microstructure, and the mathematical framework which will be used to model them.

1.2 Microstructural features

This section provides a very brief introduction to the main features of alloy microstructures that have a significant effect on a materials properties. The reader is referred to [1] for a much broader and more detailed introduction to the relationship between materials processing and microstructure.

1.2.1 Crystals and grains

In the liquid phase the different constituent elements of an alloy are typically uniformly dispersed. As the liquid cools, small solid clusters (known as **nuclei**) of 10–100 atoms form in the melt. Interatomic bonds form between the different constituent elements in the solid nuclei and cause them to adopt particular crystalline structures, e.g. body-centred cubic (bcc) or face-centred cubic (fcc). As cooling continues, **nucleation** (the formation of more nuclei) continues, as well as **growth** (the increase in size of existing nuclei). Each of the nuclei crystallites grow with their principal atomic planes in a particular orientation. This orientation is typically random. The crystallites continue to grow until they start to impinge on one another. Once all the liquid phase has gone, the resulting microstructure consists of regions with different crystallographic orientation, known as **grains**. The interfaces between the grains are known as **grain boundaries**. These internal surfaces are regions of relative disorder, where ideal crystallographic packing is not possible. These are therefore a type of **crystallographic defect**. A Voronoi tessellation is commonly used to generate a model of a representative grain structure, as shown in figure 1.2.

The simplest quantitative measure of a grain structure is the **grain size**, typically the average diameter of the grains. The grain size is small if nucleation is fast and growth is slow, i.e. many grain nuclei have time to form. Conversely, the grain size is large if the nucleation rate is slow and growth is fast, i.e. only a few grain nuclei appear before solidification completes. Both the nucleation rate and the growth rate depend on the thermodynamics (driving force) and the kinetics (atomic mobility). This is elaborated on further in chapter 7. More detailed descriptors for a grain structure can be the grain size distribution (i.e. the number of grains in a series of

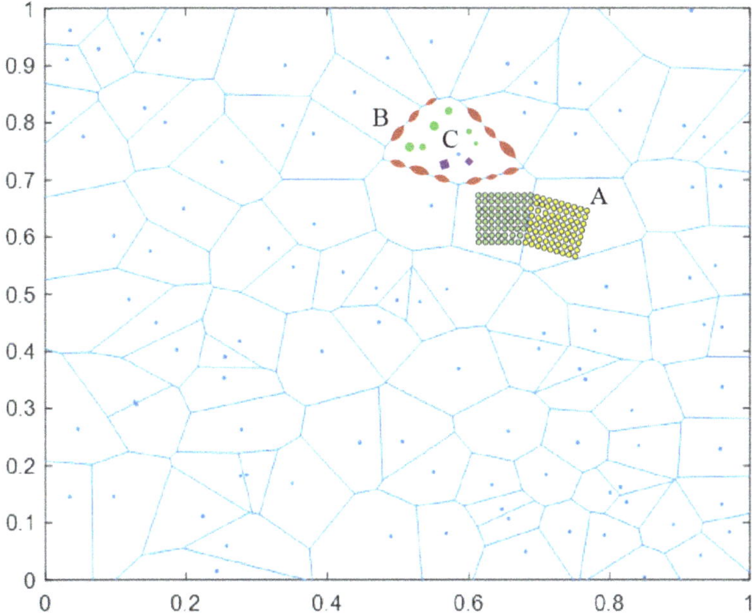

Figure 1.2. A model of a grain structure generated in MATLAB [2] using the command `voronoi(rand ([1 100]), rand([1 100]));`. This generates a Voronoi tessellation from 100 randomly distributed points. The points represent the original site of the crystallite nuclei, and the edges of the tessellation the grain boundaries. Some other microstructural features have been added manually: the grain boundary of figure 1.3(b) is superimposed on the image to illustrate the crystallography at a grain boundary (A), lenticular intergranular precipitates (red) decorate a grain boundary (B) and round (green) and facetted (purple) intragranular precipitates form within a grain (C).

particular size ranges) and, if the structure exhibits a particularly anisotropic **texture**, a grain orientation distribution (or pole figure).

1.2.2 Phases and precipitates

Alloys can have many different phases. There is typically one liquid phase (although figure 3.11(e) shows this is not always strictly true) but many different solid phases. The transition from a liquid to a solid during solidification is known as a **massive transformation** (see section 8.3 and chapter 9). In this case, one phase completely replaces another. The primary, majority solid phase that forms is referred to as the **matrix**. Note that the matrix is not always uniform in composition. Solute atoms are ejected from the solid as it grows leading to higher solute concentrations in regions of the matrix that solidify last (see section 3.2.5). Further cooling often makes the appearance of other solid phases energetically favourable. These nucleate and grow within the matrix and are known as **precipitates** if they are not a continuous phase. The schematic in figure 1.2 shows that **intergranular precipitates** prefer to form on grain boundaries and **intragranular precipitates** prefer to form within the grain. Typically precipitates make up less than 20% of the volume fraction, although in some highly structured materials, such as nickel-based superalloys, the precipitates can form over 70% of the volume. Precipitates will have a different crystal structure and/or elemental composition than the matrix. The matrix is typically ductile, whereas precipitates can often be harder and less ductile. They can therefore be valuable as a reinforcement for the matrix, increasing the yield strength. There is often a volume change associated with phase changes, meaning that there is mismatch strain between the matrix and the precipitate. This can also have a beneficial effect.

1.2.3 Defects in crystals

Defects always exist within a microstructure and are an important means by which the properties of a material can be controlled. Examples are shown in figure 1.3. These are regions of more openness in the crystal and, as such, they can be important kinetic mediators or pathways for material transport. They are also regions of higher energy density and will therefore evolve to remove themselves if a mechanism avails itself. Additionally, defects can be categorised by their dimensionality:

- Zero-dimensional defects are **point defects**. This typically refers to **vacancies**, empty lattice sites in the crystal structure, or substitutional solute atoms. Vacancies are key to facilitating the transport of material through the matrix, but their influence on the thermodynamics can generally be ignored as their concentration is very low.

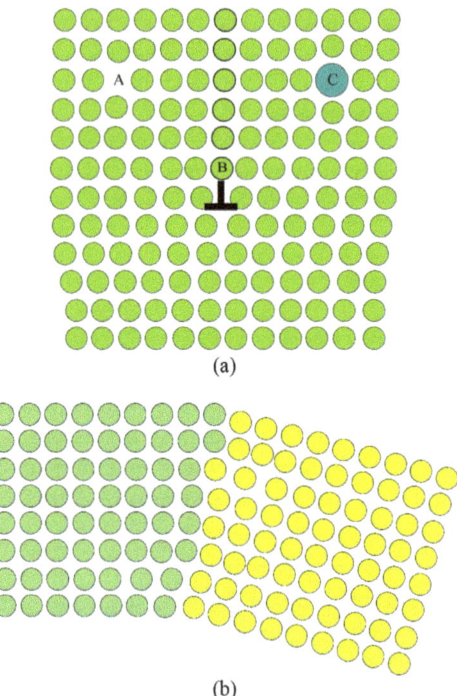

(a)

(b)

Figure 1.3. Examples of defects in crystals (a) a vacancy (A), an edge dislocation (B) and a larger substitutional solute atom (C), and (b) a grain boundary between two crystals of different orientation.

- One-dimensional defects are **line defects**. The most common of these are **dislocations**, which are a sudden change in the uniform crystallographic packing. An edge dislocation is illustrated in figure 1.3(a), where an additional column of atoms has been added above the location B. The crystal packing near to the dislocation line is highly distorted, and hence there is a long-range stress field associated with dislocations. An edge dislocation is commonly represented by the symbol shown, which represents the termination of the additional column of atoms above. To understand dislocations in three-dimensions, imagine a flat atomic surface (slip plane) within a crystal. A dislocation line forms a loop on the crystal plane. The area inside the loop is dislocated, i.e. the atoms above the plane experience a displacement (slip) of one atomic spacing relative to the atoms below the plane.
- Two-dimensional defects are **surface defects**. The most common of these are **grain boundaries.** These surround every grain, as shown in figures 1.2 and 1.3(b). They are typically 1–2 atomic spacings thick, but this dimension is very small compared to the other two-dimensions. Other common surface defects are the interfaces between precipitates and the surrounding matrix.
- Three-dimensional features would generally be discontinuous phases within another continuous phase, namely **precipitates**. Precipitates typically form preferentially at crystallographic defects such as interfaces and dislocations. As these defects have a higher energy than a regular crystal, this is energetically advantageous as the precipitate replaces the defect. Intergranular precipitates prefer to nucleate on the grain boundaries, whereas intragranular precipitates form within the grain, as illustrated in figure 1.2.

1.3 Microstructure–property relationships

The kinetics and thermodynamics of a process affect the microstructure that is generated. It is therefore useful to firstly understand how microstructural features affect the properties of a material. Determining the material properties that a given microstructure possesses is a highly complex problem which has been, and still is, the subject of much study. Here we focus on two structural mechanics properties: yield stress and creep strength. Reference [3] is recommended for a more detailed introduction. Simple analytical physics-based models are presented to allow interpretation of how different microstructural features affect these properties. However, it should be noted that these relationships are simplified, and interactions between the different mechanisms are ignored. As such, they give quantitative insight into the importance of each mechanism, but are not highly accurate where a number of mechanisms contribute simultaneously.

1.3.1 Yield stress

Some properties, such as Young's modulus and Poisson's ratio, are only weakly affected by small changes in alloy composition or processing parameters, whereas others, such as yield stress, can be greatly affected, as shown in table 1.1. The origins of this dependence are discussed here. Irreversible deformation (plasticity) in a solid

Table 1.1. Some material properties of alloys. Requested from info@grantadesign.com [4], Data courtesy of ANSYS, Inc.

Alloy	Young's modulus (GPa)	Yield stress (MPa)	Ultimate tensile strength (MPa)
High carbon steels	200–215	400–1155	550–1640
Low carbon steels	200–215	250–395	345–580
Stainless steels	189–210	170–1000	480–2240
Aluminium alloys	68–82	30–500	58–550
Copper alloys	112–148	30–500	100–550
Lead alloys	12.5–15	8–14	12–20
Magnesium alloys	42–47	70–400	185–475
Nickel alloys	190–220	70–1100	345–1200
Titanium alloys	90–120	250–1245	300–1625
Zinc alloys	68–95	80–450	135–520

is mediated by the movement of dislocations. A material is strengthened by incorporating obstacles into the microstructure that prevent the motion of dislocations, namely large solute atoms, other dislocations, precipitates and grain boundaries. The contribution of each of these is considered further below.

There are essentially two types of obstacles in a microstructure: those that a dislocation can bypass (by a mechanism known as **Orowan bowing**) and those that cannot be bypassed. Most 0D (solute), 1D (dislocation) and 3D (precipitate) obstacles fall into the by-passable category. 2D (grain boundary) obstacles can entirely enclose a dislocation and hence cannot be bypassed by Orowan bowing. The contribution from by-passable obstacles is considered first.

Obstacles can pin segments of dislocation line but additional stress can allow the dislocation to bypass the obstacles, as shown in figure 1.4. A dislocation line has an energy of $\frac{1}{2}Gb^2$ per unit length, where G is the shear modulus of the matrix and $b \approx$ 0.25 nm is the slip accommodated by the dislocation (Burgers vector). As such, work needs to be done (by the applied force) to increase the length of a dislocation line.

The critical shear stress required to extend the dislocation line sufficiently for escape can be calculated by equating the work done by the applied load to the energy required to extend the dislocation line around the obstacle. This gives the Orowan stress

$$\tau = \beta \frac{Gb}{L}$$

where L is the distance between obstacles and β is a constant strengthening factor. This shows that a large number of small obstacles which are close together (small L) is more effective at blocking dislocation movement than a small number of large obstacles which are far apart (large L). Some types of obstacle are more effective at pinning dislocation segments than others, and hence each type has a different strengthening factor:

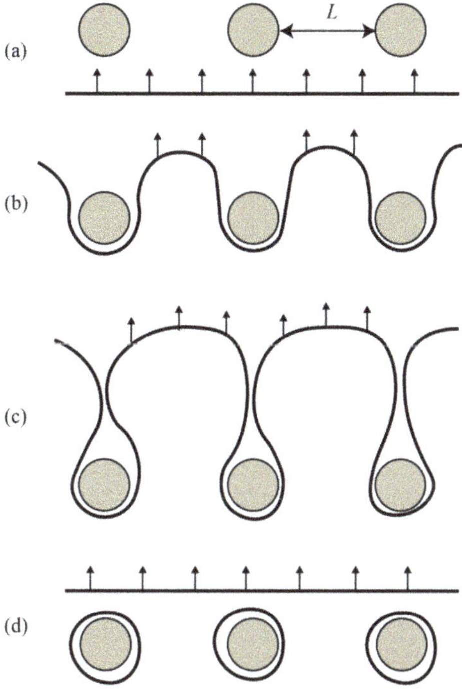

Figure 1.4. Schematic of Orowan bowing mechanism. (a) A straight dislocation line approaches a number of precipitate obstacles, (b) an increase in the force on the dislocation drives it between the precipitates increasing the length of the dislocation, (c) additional force sees the dislocations almost touching on the other side, (d) the dislocations have touched leaving a dislocation around each precipitate. The straight dislocation line glides on to the next set of obstacles.

- **Solid solution strengthening** (denoted by subscript S) arises from solute atoms (0D obstacles), which locally distort the matrix if they are smaller or larger than their surrounding matrix atoms, as shown in figure 1.3(a). They are typically fairly weak obstacles ($\beta_S = 0.01 - 0.05$ depending on the size mismatch of the substitutional atom), although the spacing between them, $L_S = \dfrac{b}{c^n}$, can be very small. As such they can still produce a substantial contribution. Here c is the molar concentration of the solute, where $n = 1/2$ for dilute concentrations and $n = 2/3$ for higher concentrations due to strain field interactions.

- **Forest hardening** (denoted by subscript F) arises from dislocations (1D obstacles). The simplest measure of the dislocation state is the dislocation density ρ (m^{-2}) which is the total dislocation line length per unit volume. The dislocation density can increase due to plastic strain (work hardening). Typically $10^{10} < \rho < 10^{15}$ m^{-2}. Dislocations are moderately effective obstacles ($\beta_F = 0.1 - 0.3$) with an average spacing of $L_F = \dfrac{1}{\sqrt{\rho}}$.

- **Precipitate hardening** (denoted by subscript P) arises from precipitate (3D) obstacles, which are some of the most effective obstacles ($\beta_P = 0.5 - 0.8$). If the average precipitate radius is r_p and the precipitate volume fraction is f then the average spacing between them is $L_P = 1.6 r_p \left(\sqrt{\dfrac{\pi}{4f}} - 1 \right)$. In general a fine dispersion of small precipitates is desired for optimal yield stress, although it should be noted that if the precipitates are too small (<20–50 nm) then the dislocations can avoid bypassing them by cutting through them.
- As mentioned previously, **grain boundary strengthening** (denoted by subscript G) due to grain boundaries (2D obstacles) is a different case, as grain boundaries cannot be bypassed by Orowan bowing. Orowan showed that the total plastic shear strain due to the motion of dislocations is necessarily $\gamma_p = b\rho\bar{x}$, where $\bar{x}$ is the mean distance travelled by each dislocation. In polycrystals it is assumed that dislocations can travel a distance $\bar{x} \approx d$ before they are pinned at a grain boundary, where d is the grain size (diameter). More dislocations must therefore be generated within each grain to mediate increases in the plastic strain, such that $\rho = \dfrac{\gamma_p}{bd}$. Forest hardening occurs as the dislocation density increases, yielding $\beta_G \approx \beta_F$ and $L_G \approx \dfrac{1}{\sqrt{\rho}} = \sqrt{\dfrac{bd}{\gamma_p}}$.

To first order, the total yield stress of a microstructure can be approximated as the sum of these contributions

$$\tau_Y = \tau_0 + \frac{Gb}{L_{\text{eff}}}$$

where τ_0 is the intrinsic strength of the matrix crystal (usually small), and the effective length of the microstructure is a weighted combination of the many length scales from the different microstructural features

$$\frac{1}{L_{\text{eff}}} = \frac{\beta_S}{L_s} + \frac{\beta_F}{L_F} + \frac{\beta_P}{L_p} + \frac{\beta_G}{L_G}.$$

The second term on the right-hand side gives the standard Taylor model for forest hardening $\left(\tau_F \propto \sqrt{\rho} \right)$ and the third term gives the classic Hall–Petch relation $\left(\tau_G \propto 1/\sqrt{d} \right)$ for grain-size strengthening. Higher order interactions between the contributions are ignored here, but can be significant if any of the terms are of similar magnitude.

1.3.2 Creep strength

The Hall–Petch relation demonstrates that a smaller grain size leads to a higher yield stress, which can be considered to be a positive outcome for most metals if they are operating below half their melting temperature. However, at higher temperatures, a material can deform slowly over time by material rearrangement by either diffusion (diffusion creep) or dislocation climb (power law creep). This can be a major problem in high temperature applications, such as steam turbines (used for power generation) and gas turbines (aircraft jet engines), as the material can fail by creep after a long service history (possibly up to 25 years or more in a steam turbine).

Diffusion creep is accelerated by diffusion along grain boundaries (Coble creep) or through the matrix from one grain boundary to another (Nabarro–Herring creep). The irreversible creep strain rate for diffusion creep is strongly influenced by the microstructural length scale due to the grain size, d, where typically

$$\frac{d\varepsilon_c}{dt} = \frac{A(T)}{d^n}$$

where the material parameter $A(T)$ is a function of temperature T and proportional to the dominant kinetic rate constant, and n is a creep exponent. For Nabarro Herring creep $n = 2$ and for Coble creep $n = 3$. These both show that diffusional creep is most rapid when the grain size is small, in direct contrast to the requirements for a high yield stress. These mechanisms dominate at low stresses, with Coble creep more prevalent at lower temperatures, and Nabarro–Herring creep becoming more influential at higher temperatures.

At high stresses and temperatures, **power law creep** dominates. This occurs by increased dislocation movement. Under the action of a shear stress, dislocations will rapidly **glide** along slip planes until they meet an obstacle. At low temperatures, the dislocation will only be able to bypass the obstacle if the stress is increased to facilitate Orowan bowing. However, at higher temperatures, dislocations can bypass an obstacle by **climbing** slowly over the obstacle, moving from one slip plane to another. For this to occur a vacancy must diffuse into the dislocation. For instance, if the vacancy (A) exchanges with the atom at the dislocation core (B) in figure 1.3(a) then the dislocation effectively moves up one atomic spacing. As the temperature increases, the number and mobility of vacancies also rises, facilitating the climb process and rendering obstacles less effective at pinning dislocations. Power law creep models have largely been phenomenological in nature, although physics-based models that are related to microstructural quantities are more powerful as they can still predict the creep response over a range of alloys and microstructures. The Dyson model [5] for nickel-based superalloys has had some success in this regard

$$\frac{d\varepsilon_c}{dt} = \frac{\rho(r_P + L_P)(1 - f)D_s}{3r_P} \sinh\left(\frac{b^2 L_p}{3k_B T}(\sigma_e - \sigma_b)\right)$$

for $\sigma_e > \sigma_b$, where the new additional parameters are the diffusion coefficient D_s, Boltzmann's constant k_B, the applied (effective) stress σ_e and the critical back stress σ_b that must be surmounted for creep to initiate (this is related to the Orowan stress due to obstacles previously discussed). The model shows that, although climb of dislocations cannot be prevented at high temperatures, the introduction of obstacles through precipitation hardening and solute strengthening can help arrest this process. Increasing the dislocation density by work hardening may increase the yield stress but it can also increase the creep rate as power law creep is mediated by their movement.

1.4 A framework for microstructure evolution

The rest of this book concerns itself with the mathematical tools required for the prediction of how microstructures form and change over time. The first part introduces the essential concepts of phase thermodynamics, mainly focusing on equilibrium states. This will also show how the chemical driving force for microstructural change is determined. The second part on kinetics adds space and time to the model by including mass transport. Various aspects of microstructural evolution are then considered based on the foundations of thermodynamic and kinetic modelling.

1.4.1 PANDAT software and TDB databases

There are a number of commercial thermodynamic packages that could be used to demonstrate the concepts discussed in this book, namely PANDAT [6], Thermocalc [7] and JMatPro [8]. There is also the OpenCalphad opensource project [9]. The free demonstration version, PANDAT 2020 DEMO, is used here as it is easily available for students and lecturers to download [6] and follow the examples in the following chapters, as well as the many examples provided by the software provider Computherm. The demo version of PANDAT is fully functional, except that it is limited to 3 components or less. This allows binary and ternary phase diagrams to be calculated and as such is perfectly adequate for introducing the concepts required here.

Thermodynamic data for a particular material is defined in a TDB (Thermodynamics DataBase) file. This uses a standard format to define the thermodynamics of different phases as a function of composition and temperature. For most of the work presented here, simple TDB files are created that represent idealised systems. This is to help the reader appreciate the importance of the few terms required, and demonstrate the wealth of variety of behaviour that can be achieved by varying only one or two key parameters. The development of (encoded) TDB files for real, multi-component systems is a lengthy and costly process, and hence they are not generally freely available. However, there are a number of free resources on the web for TDB files which are listed at [10]. The Japanese National Institute for Materials Science (NIMS) website is especially useful for a wide range of binary and ternary phase diagrams [11]. It is free but you must register to use it first.

To follow the examples provided in this book, you will need to set up PANDAT and download a TDB file using these instructions.

PANDAT: installation and sample database download
1. Download and install the free demo version of Pandat from [6].
2. Register and sign in at the MatNavi NIMS thermodynamic database web site [11].
3. You will be presented with a periodic table as shown in figure 1.5. The blue highlighted elements are those with binary database files for download.

Computational Phase Diagram Database

1	2	3	4	5	6	7	8	1	2	3	4	5	6	7	0
H 1															He 2
Li 3	Be 4									B 5	C 6	N 7	O 8	F 9	Ne 10
Na 11	Mg 12									Al 13	Si 14	P 15	S 16	Cl 17	Ar 18
K 19	Ca 20	Sc 21	Ti 22	V 23	Cr 24	Mn 25	Fe 26	Co 27	Ni 28	Cu 29	Zn 30	Ga 31	Ge 32	As 33	Se 34	Br 35	Kr 36
Rb 37	Sr 38	Y 39	Zr 40	Nb 41	Mo 42	Tc 43	Ru 44	Rh 45	Pd 46	Ag 47	Cd 48	In 49	Sn 50	Sb 51	Te 52	I 53	Xe 54
Cs 55	Ba 56	Lanthanide 57-71	Hf 72	Ta 73	W 74	Re 75	Os 76	Ir 77	Pt 78	Au 79	Hg 80	Tl 81	Pb 82	Bi 83	Po 84	At 85	Rn 86
Fr 87	Ra 88	Actinide 89-103													

| | | La 57 | Ce 58 | Pr 59 | Nd 60 | Pm 61 | Sm 62 | Eu 63 | Gd 64 | Tb 65 | Dy 66 | Ho 67 | Er 68 | Tm 69 | Yb 70 | Lu 71 | |
| | | Ac 89 | Th 90 | Pa 91 | U 92 | Np 93 | Pu 94 | Am 95 | Cm 96 | Bk 97 | Cf 98 | Es 99 | Fm 100 | Md 101 | No 102 | Lr 103 | |

Figure 1.5. The home screen of the MatNavi NIMS thermodynamic database web site [11] (reproduced with permission).

4. Click on 'Al(13)' to see which binaries are available for aluminium. Note that files do not exist for all binary systems. For all 103 elements that would result in 5253 combinations.

5. There are about 40 binary systems where Al is one of the two elements. (For reference, note that if you cannot find the particular system you are interested in, it is possible that it has been characterised in the literature but it is not in the NIMS database, e.g. the H–Zr system is not in the NIMS database but the thermodynamic parameters that characterise the system have been published [12]. You would have to create the TDB (thermodynamic database) file manually though. This is not as daunting a task as it might seem and the examples in this book will help you understand how the TDB files are constructed.)

6. We are only interested in pure aluminium at this point, so it does not matter which binary system we choose. The aluminium–silicon system is relatively simple so we shall use that. Click on the link for 'Al–Si'.

7. You will see a picture of the phase diagram that the most up-to-date TDB file produces. There are typically a number of options for the choice of TDB file, as each binary database has been improved by further research and data over the years. For Al–Si there are three files, from 1984, 1986 and 1998. The top option (1998) is always the newest and usually the best to download. The journal

> reference for the paper where the data is published is given for reference. Click on the 'TDB File' link that is part of the 1998COST reference.
>
> 8. The contents of the TDB file should appear in a new page. Press CTRL-A and CTRL-C to copy the contents into the clipboard. Open a simple word processing package (e.g. Notepad in Windows) and press CTRL-V to paste the file into the editor. Save the file as AlSi.tdb. (In Notepad you need to switch 'Save As Type' to 'All files' otherwise it will add an extra '.txt' extension to a text file).

1.4.2 A mathematical framework

A kinetic variational principle is used in Part II of this book as a consistent mathematical framework within which underlying microstructural evolution models can be derived from basic principles. The advantage of using this method is that the evolution equations do not have to be solved exactly. An approximate form for the solution can be proposed with a number of degrees of freedom (such as the precipitate radius or the grain size). A variational functional is then formed and minimised with respect to the degrees of freedom to find the most optimal kinetic pathway for the system given the approximations made. The process has similarities to other variational methods in this respect, such as the finite element method. This makes the derivations more physical and less mathematically complex. The important scaling relationships are obtained, although precise geometric factors are only approximated. The latter are not considered to be too important, as the geometric factors are generally of the order of unity, and they are typically strongly coupled with kinetic rate constants, which are generally not known to a high degree of accuracy.

The general variational functional can be written as

$$\Pi = \Psi + \dot{G} \tag{1.1}$$

which is minimised when it is stationary

$$\delta\Pi = 0. \tag{1.2}$$

Here Ψ represents the kinetics (rate of energy dissipation due to the kinetic processes) and $\dot{G}$ represents the thermodynamics (the rate of change of Gibbs free energy). This latter term provides the energetic driving force for a microstructure to change and is non-positive ($\dot{G} \leqslant 0$). This shows that microstructures always evolve to reduce their Gibbs free energy and will only stop once a minimum energy state (equilibrium) is reached ($\dot{G} = 0$). However, in general, a microstructure never reaches its minimum energy state. The chemical (phase) contribution to the Gibbs free energy is determined in Part I. The different kinetic processes and their rates of dissipation are determined in Part II.

References

[1] Ashby M F and Jones D R H 2012 *Engineering Materials 2: An Introduction to Microstructures, Processing, and Design* 4th edn (Oxford: Pergamon)

[2] MATLAB R 2020 Copyright 1984–2018 The MathWorks, Inc

[3] Ashby M F 2012 *Engineering Materials 1: An Introduction to Properties, Applications, and Design* 4th edn (Oxford: Pergamon)

[4] Cambridge University Engineering Department databook www-mdp.eng.cam.ac.uk/web/library/enginfo/cueddatabooks/materials.pdf [accessed 1 October 2021]

[5] Dyson B 2009 Microstructure based creep constitutive model for precipitation strengthened alloys: Theory and application *Mater. Sci. Technol.* **25** 213–20

[6] PANDAT software: http://e.informer.com/computherm.com/ [accessed 1 October 2021]

[7] Thermocalc software: https://thermocalc.com/ [accessed 1 October 2021]

[8] JMatPro software: www.sentesoftware.co.uk/jmatpro [accessed 1 October 2021]

[9] OpenCALPHAD software: www.opencalphad.com/ [accessed 1 October 2021]

[10] OpenCALPHAD databases: http://opencalphad.com/databases.html [accessed 1 October 2021]

[11] National Institute for Materials Science (NIMS) database: http://cpddb.nims.go.jp/cpddb/periodic.htm [accessed 1 October 2021]

[12] Konigsbergera E, Eriksson G and Oates W A 2000 Optimisation of the thermodynamic properties of the Ti–H and Zr–H systems *J. Alloys Compd.* **299** 148–52

Part I

Thermodynamics and phase diagrams

IOP Publishing

Thermodynamics, Kinetics and Microstructure Modelling

Simon P A Gill

Chapter 2

Thermodynamic quantities

The thermodynamics of phases are defined by a quantity known as the Gibb's free energy. This is defined as

$$G = H - ST \tag{2.1}$$

where H is enthalpy, T is temperature (in Kelvin) and S is entropy.

There are many excellent textbooks on thermodynamics, phase diagrams and phase transformations in alloys. The reader is referred to [1] for a more detailed background. The aim of this book is to introduce the thermodynamics of phase change through a number of practical examples and exercises to aid understanding. To this end, enthalpy, entropy and Gibbs free energy are introduced in the following subsections in the context of pure aluminium as it is heated from 400 °C to 800 °C and undergoes a solid-to-liquid transition (melting) at 660 °C.

As introduced in section 1.4.1, this book utilises the free demo version of the PANDAT software for generating binary and tertiary phase diagrams and understanding the physics behind it. Hence it is highly recommended that readers download and use PANDAT as they work through the next four chapters. Learning through hands-on experience is always more stimulating and informative.

2.1 Enthalpy

The enthalpy is the sum of the internal energy of the substance, U, and the product of the pressure P and the volume V

$$H = U + PV. \tag{2.2}$$

It represents the amount of energy that is transferred into or out of the system. In this context this energy is typically in the form of heat (thermal energy) as the PV term is small for solids and liquids. The internal energy U is the sum of the energy of the bonds between atoms plus their kinetic (thermal vibrational) energy. To illustrate this, use PANDAT to plot the change in the enthalpy of aluminium as it is heated from a solid to a liquid.

doi:10.1088/978-0-7503-3147-0ch2

PANDAT: plotting enthalpy

1. Follow the setup instructions in section 1.4.1 if you have not already installed Pandat and downloaded the TDB file AlSi.tdb.
2. Run Pandat. Under 'Create a New Workspace', select the 'PanPhaseDiagram' icon and click the 'Create' button.
3. Select 'Databases → Load TDB or PDB (Encrypted TDB)' and choose the file 'AlSi.tdb'.
4. Under 'Select Components', click on the element 'Al' and press the central blue arrow pointing to the right to move it from the 'Available Components' column to the 'Selected Components' column. Press 'OK'.
5. Select 'PanPhaseDiagram → Line Calculation'.
6. The 'Start Point' represents the temperature and composition on the left-hand side of the plot, and the 'End Point' the temperature and composition on the right-hand side of the plot. Change the 'Start Point' value of 'T[C]' to 400 and the 'End Point' value of 'T[C]' to 800. The composition of the alloy 'x(Al)' must remain at 1 as we are interested in the thermodynamics of pure Al only.
7. Click on 'Options' and select 'Table' in the left-hand column. We wish to plot the enthalpy 'H'. Pandat creates a table of values that can be plotted. To add enthalpy to the table, check the checkbox by the symbol 'H' in the central 'Choose table Columns' column. This should then appear as an option in the right-hand 'Choose Y-axis Properties' columns. Check the checkbox against 'H' in this column as well, and uncheck any other boxes that are checked, e.g. 'f(@*)'. Press 'OK' to exit 'Options'.
8. Press the 'OK' button to complete the plot, shown in figure 2.1. Note that, as shown, you can add arrows and text to your plot using the 'Graph' options.

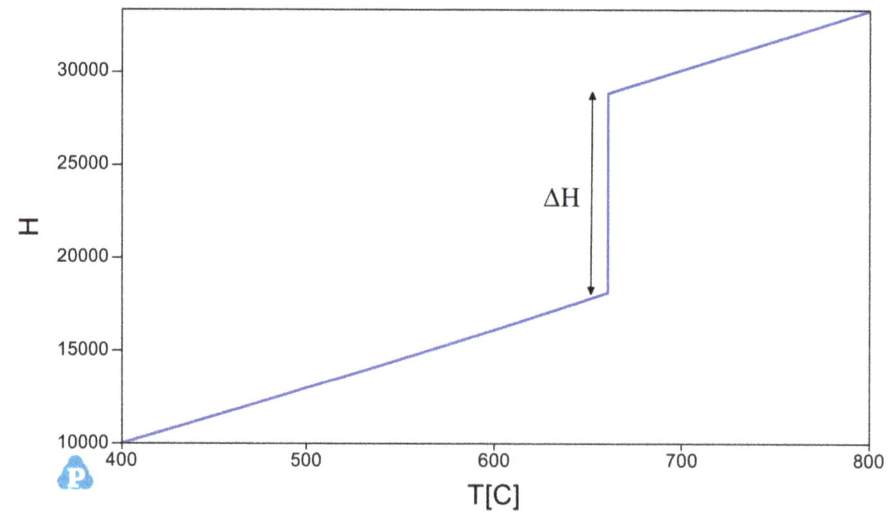

Figure 2.1. Enthalpy of Al as a function of temperature T. Created using PANDAT [3].

Figure 2.1 shows the change in the enthalpy of Al as it is heated from a solid at 400 °C. The enthalpy increases steadily as it absorbs more heat. At the melting temperature of 660 °C, there is a jump in the enthalpy, ΔH. This shows that a substantial amount of heat is required to increase the temperature beyond the melting point. This is called the **enthalpy of fusion** or **latent heat of fusion**, and commonly denoted L_f. Further heating continues to raise the temperature of the (now liquid) Al. The units of enthalpy on the vertical axis are J mol^{-1}.

The amount of thermal energy required to raise one mole of a substance by one degree Kelvin at constant pressure is called the **specific heat capacity**, and is defined in terms of the enthalpy as

$$c_p = \left(\frac{\partial H}{\partial T}\right)_P \tag{2.3}$$

where the P denotes constant pressure is maintained. This is simply the slope of the enthalpy diagram in figure 2.1. Plot this using PANDAT.

PANDAT: plotting specific heat capacity

1. Run another 'Line Calculation'. Note you can use the [—] icon if you wish.

2. Select 'Options' and check the 'Cp' checkbox in the central 'Choose table Columns' column. Check the 'Cp' checkbox in the right-hand 'Choose Y-axis Properties' column to plot this and uncheck the 'H' checkbox to remove this from the plot. Press the 'OK' button, and then press 'OK' again.

Figure 2.2 shows that the heat capacity in each phase is roughly constant in the liquid phase (above 660 °C) and a weak function of temperature in the solid phase (below 660 °C). Note that alloys, which have more than one element, melt (and

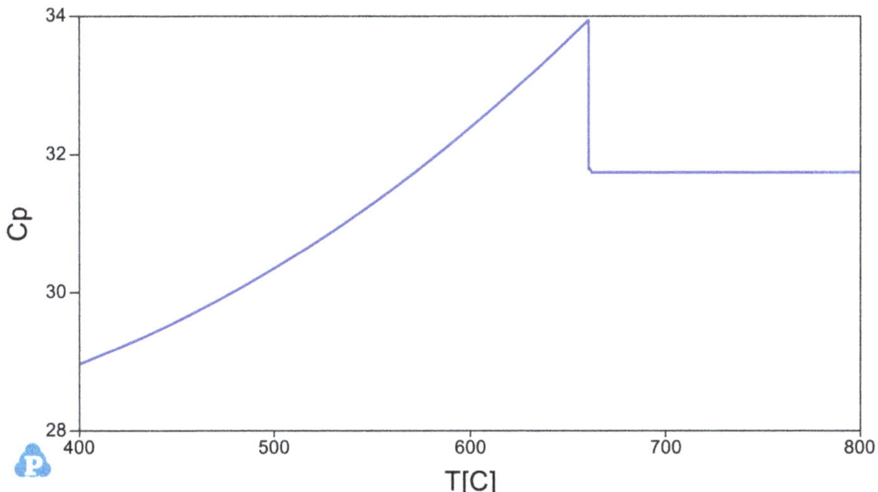

Figure 2.2. Specific heat capacity of Al as a function of temperature T. Created using PANDAT [3].

solidify) over a temperature range. Only pure metals and alloys of a specific composition (known as the eutectic composition) melt at a single temperature.

2.2 Entropy

Entropy is a measure of the number of ways that the atoms in a crystal can be arranged. The larger the entropy, the more permutations a given crystal has. The second law of thermodynamics states that systems always act to increase their entropy, i.e. adopt a state that has the highest number of similar permutations and hence is the state that is most likely to occur based on basic statistical mechanics. There are many different types of entropy, but in this context we are concerned with the **entropy of mixing**.

PANDAT: plotting entropy
1. Run another 'Line Calculation' and see if you can plot the entropy 'S' without instructions.
2. If you need help, select 'Options' and check the 'S' checkbox in the central 'Choose table Columns' column. Check the 'S' checkbox in the right-hand 'Choose Y-axis Properties' column to plot this and uncheck the 'Cp' checkbox to remove this from the plot. Press the 'OK' button, and then press 'OK' again.

The change in entropy as Al changes from a solid to a liquid is shown in figure 2.3. As seen in the enthalpy plot, there is also a discontinuity at the melting point, ΔS. This represents the change in order during the phase transition. The

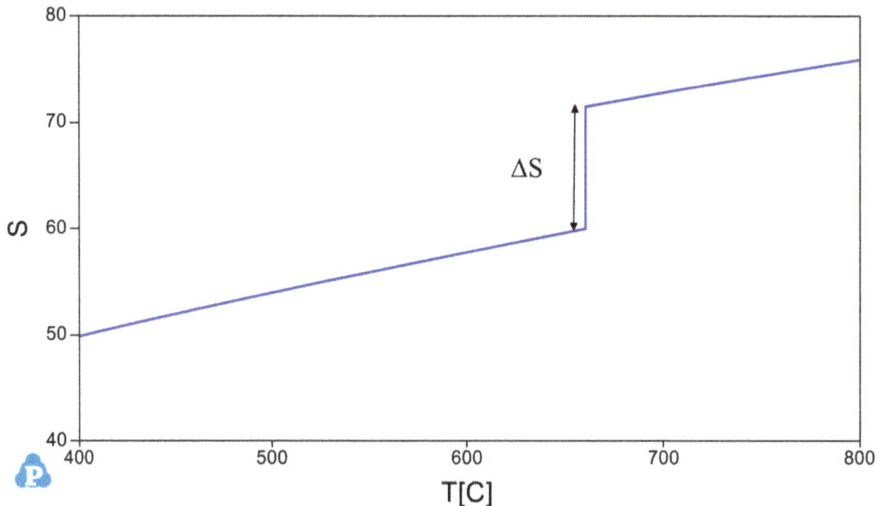

Figure 2.3. Entropy of Al as a function of temperature T. Created using PANDAT [3].

crystalline solid has a high degree of order (low entropy) whereas the liquid phase is much less ordered (high entropy). The units of entropy are $J \; mol^{-1} \; K^{-1}$.

From a modelling perspective it is interesting to note that entropy is a theoretical construct required for continuum mechanics. In atomistic models, such as molecular dynamics simulations, enthalpy is introduced through the interatomic potentials, but entropy does not need to be included explicitly as the atoms will naturally rearrange to form a high probability state.

2.3 Gibb's free energy

In general, systems will evolve (if possible) to minimise their Gibbs free energy. The Gibb's free energy, defined by (2.1), combines the enthalpy and entropy contributions of figures 2.1 and 2.2. To see this plot the total Gibb's free energy using PANDAT.

PANDAT: plotting the total Gibb's free energy
1. Following the same procedure as before, run a 'Line Calculation' to plot the parameter 'G'.

Figure 2.4(a) shows that the discontinuous enthalpy and entropy contributions of figures 2.1 and 2.2 produce a continuous function for the Gibb's free energy. This demonstrates that the additional thermal energy (enthalpy of fusion) supplied at the melting temperature T_M is precisely that required to increase the entropy (disorder) of the solid when it becomes a liquid, such that

$$\Delta H = T_M \Delta S = L_f. \tag{2.4}$$

Figure 2.4(a) shows the minimum Gibbs free energy of the entire system for 1 mole of Al. The curve is continuous but there is a sudden change in slope at the melting point of 660 °C. To see why this is, plot the energy of the liquid and solid phases separately.

PANDAT: plotting the Gibb's free energy of all phases
1. Following the same procedure as before, run a 'Line Calculation' and select the parameter '$G(@*)$' to plot in the 'Option'. The '@' refers to an individual phase and the '*' means include all the phases. Press 'OK' to exit 'Options'.
2. In 'Line Calculation', click on the 'Select Phases' button. The AlSi binary system includes a phase called 'DIAMOND_A4' which does not exist in pure Al. To remove this phase from the plot, click the two central blue arrows pointing to the left to move 'DIAMOND_A4' from the 'Entered Phases' to the 'Suspended Phases' column. Press 'OK'.
3. At the bottom of the 'Line Calculation' window, check the 'Individual Phases' checkbox. This ensures that the energy of a phase is plotted even when it is not the equilibrium phase. (Try plotting the energy without this button checked and you can see the difference for yourself.) Press 'OK' to produce the plot.

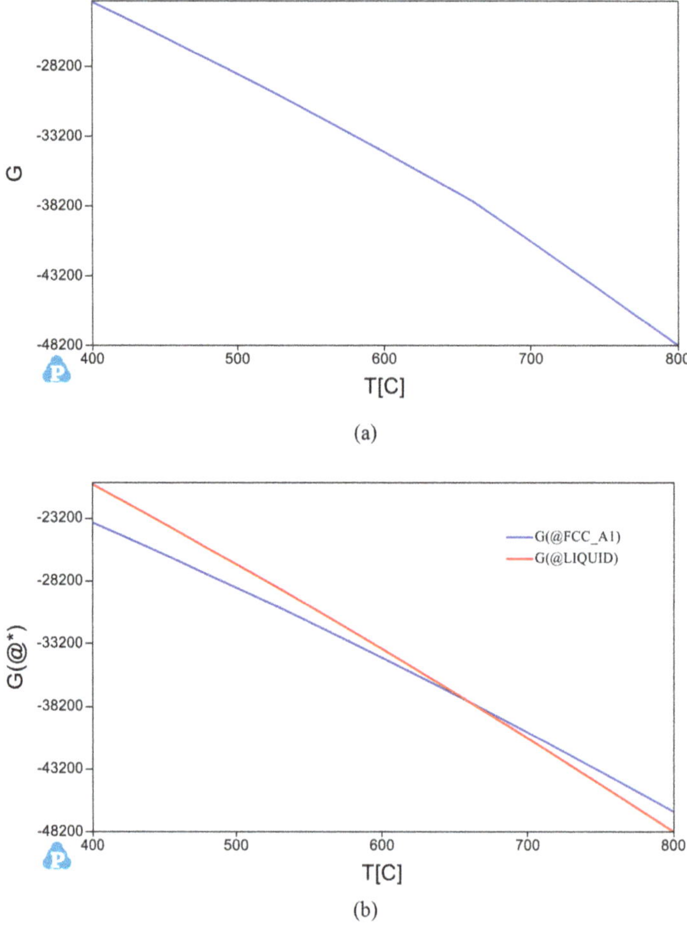

Figure 2.4. (a) The minimum (equilibrium) Gibbs free energy of the whole system, and (b) the individual Gibbs free energies of the liquid and solid phases of Al as a function of temperature T. Created using PANDAT [3].

The Gibbs free energy of the two phases of Al (liquid and solid) are shown in figure 2.4(b). Notice that both functions are continuous in slope as well as value. The melting point of 660 °C coincides with the temperature at which the Gibbs free energy of the liquid becomes less than that of the solid. Comparison with figure 2.4(a) shows that the system adopts the energy of the solid phase below the melting temperature and the energy of the liquid phase above it. This is the origin of the change in slope of the curve at the melting point. The solid phase is denoted FCC_A1 which describes the particular face-centred cubic (FCC) crystal structure of the solid phase using the standard crystallographic nomenclature.

The Gibbs free energy is expressed in the following near polynomial form

$$G(T) = cT \ln(T) + a_0 + a_1 T + a_2 T^2 + \cdots + a_{-1} T^{-1} + a_{-2} T^{-2} + \cdots \qquad (2.5)$$

where c and the a_n are constants, with the integer typically in the range $-9 \leqslant n \leqslant 9$, and roughly 3–6 of these terms being non-zero in total. The $T\ln(T)$ term arises for the need to obtain a constant value for the heat capacity, which has the alternative definition to (2.3) of $c_P = -T\dfrac{\partial^2 G}{\partial T^2}$. The material constants for the Gibb's free energy are obtained from numerical fits to empirical data. The data is typically obtained by cooling a sample from a liquid and measuring the heat that is released during the solidification process to calculate the enthalpy change. A differential scanning calorimeter is used to measure the heat released accurately. The entropy change is known from theory (see section 3.1.1) and hence $G(T)$ can be determined. The reader is referred to [2] for a nice illustration of the calibration process.

2.4 Phase diagrams for pure substances

So far the thermodynamics of aluminium have been used to introduce the fundamental concepts of phase energy in a unary (one component) system. In the next chapter we will look at the thermodynamic of binary (two component) systems. To do this a fictitious, idealised material is constructed to demonstrate the key features of thermodynamic expressions in the simplest form possible.

The idealised material will have two phases: the SOLID phase and the LIQUID phase. The total Gibb's free energy (J) will be denoted by $G(T)$ and the Gibb's free energies per mole (J mol^{-1}) for each phase are denoted by the lower case symbols $g_S(T)$ and $g_L(T)$ for the solid and liquid phases respectively. The melting temperature, T_M, demarcates the phase boundary. These are illustrated on figure 2.5.

At equilibrium the lowest energy phase will be observed. For $T < T_M$ we have $g_S < g_L$ so the material is SOLID. For $T > T_M$ we have $g_L < g_S$ so the material is

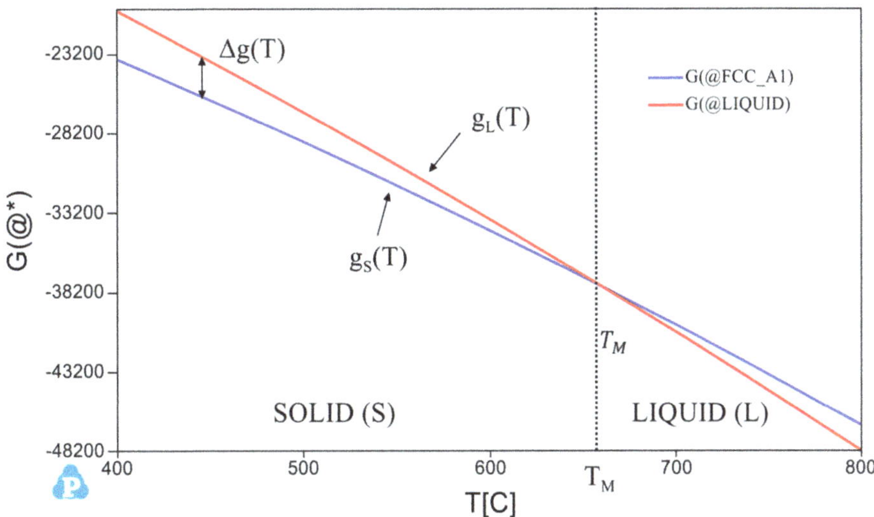

Figure 2.5. A copy of figure 2.4 labelling the phase energies and liquid/solid phase boundary. Created using PANDAT [3].

LIQUID. The minimum energy of the system therefore switches from the SOLID (blue) curve to the LIQUID (red) curve above the melting point. Consider the difference in the Gibbs free energy between the two phases as the SOLID phase changes to the LIQUID phase, $\Delta g_{S \to L}(T) = g_L - g_S$. This is the important quantity in determining which phase is present: SOLID if $\Delta g_{S \to L} > 0$ and LIQUID if $\Delta g_{S \to L} < 0$. When $\Delta g_{S \to L}(T) = 0$ we are at a phase boundary

$$\Delta g_{S \to L}(T_M) = \Delta H - T_M \Delta S = 0$$

which gives equation (2.4). The simplest functional form we can use to describe the energy difference between the two phases is therefore

$$\Delta g_{S \to L}(T) = \Delta S(T_M - T) \tag{2.6}$$

where equation (2.4) defines the constant $\Delta S = \dfrac{L_f}{T_M}$. This expression is adequate to define a phase transition in a two phase system with one (pure) component. Two component systems are considered in the next chapter, where versions of equation (2.6) are used to describe the melting transition of each of the two pure components.

References

[1] Porter D A, Easterling K E and Sherif M Y 2018 *Phase Transformations In Metals And Alloys* 3rd edn (Boca Raton, FL: CRC Press)
[2] Dissemination of IT for the Promotion of Materials Science (DoITPoMS) website: https://doitpoms.ac.uk/tlplib/phase-diagrams/cooling.php [accessed 1 October 2021]
[3] Pandat: software suite for thermodynamic calculation and kinetic simulation of multi-component alloys. CompuTherm LLC, Madison, Wisconsin, USA www.computherm.com [accessed 1 October 2021]

IOP Publishing

Thermodynamics, Kinetics and Microstructure Modelling

Simon P A Gill

Chapter 3

Binary systems

The development of an ideal, fictitious material is now extended to model a binary system, containing two elements. This generates a much richer variety of material behaviour which is reflected in the increased complexity of the phase diagrams. This introduces the concept of entropy, which models the expected randomness that is seen in nature, as well as the lever rule for determining the amount of each phase in a two phase region, the concept of solute partitioning as a liquid cools to a solid over a temperature range, and how to deal with stoichiometric phases.

The binary system contains two elements, which we label A and B. Let the number of A atoms be N_A, the number of B atoms N_B and the number of atomic sites in the crystal lattice N_{sites}. A number N_{Va} of the lattice sites can be occupied by vacancies (i.e. unoccupied) such that the total number of lattice sites is $N_{sites} = N_A + N_B + N_{Va}$. The **atomic fraction** (at%) of each element is therefore

$$x_A = \frac{N_A}{N_{sites}} \qquad x_B = \frac{N_B}{N_{sites}} \qquad x_{Va} = \frac{N_{Va}}{N_{sites}}.$$

Vacancies are simply included as another element in the thermodynamic model, i.e. it is treated as a three component system. It is easy to see that the atomic fractions satisfy $x_A + x_B + x_{Va} = 1$. The contribution from vacancies is typically very small (i.e. $x_{Va} \ll 1$) so it is assumed here that all the lattice sites in the crystal are occupied and that

$$x_A + x_B = 1. \tag{3.1}$$

The concentrations of A and B are therefore not independent, e.g. if the amount of A is 40 at% then we know that the amount of B must be 60 at%. In practice thermodynamic databases represent compositions by **molar fraction** (mol%). This is similar to the atomic fraction but uses the number of molecules, n_A, of a component rather than the number of atoms, N_A. The molar fraction and atomic fraction are the same for elemental compositions, where an atom is considered to be the constituent molecule, and hence the atomic fractions (x_A etc) are also the mole fractions. The difference between atomic and mole fraction only arises for particular phases where the atoms combine in strict ratios, e.g. water, where the water molecule is H_2O and hence $n_{H_2O} = 2N_H + N_O$. This will be discussed in further detail in section 3.3.

Note that the elemental compositions of alloys are often given by a different measure, **weight fraction** (wt%). For a general M component material system, the weight fraction of component i is

$$X_i = \frac{w_i n_i}{\sum_{j=1}^{M} w_j n_j}$$

where w_i is the weight per mole of component i. Interpretation of visual images of microstructure will generally show components (and phases) by area fraction which is most closely related to **volume fraction** (vol%)

$$v_i = \frac{\Omega_i n_i}{\sum_{j=1}^{M} \Omega_j n_j}$$

where Ω_i is the molar volume (volume per mole) of component i. In general the atomic fraction is most suitable for modelling material transport within a microstructure as the number of atoms is always conserved. Molar volume and mole number are not necessarily invariant during phase changes.

The molar Gibbs free energy of a multi-component phase, $g(x_i, T)$, can now be expressed as a function of temperature T and the molar fraction of its components, x_i, where $i = A$ or B in this context. The rest of this chapter considers the different forms this function can take and how it influences the equilibrium properties of multiphase binary systems.

3.1 A single phase system with two components

A single phase is considered, which we call SOLID, although the energetics still apply for liquid phases. Gas phases, which are strongly affected by pressure and volume change, are not considered here. As seen in equation (2.1), the two main contributions to the Gibbs free energy are enthalpy and entropy. These are introduced for binary systems in the following sections.

3.1.1 Entropy of mixing

The entropy of an ideal solution acts to drive any phase towards its most likely atomic arrangement, i.e. randomly mixed. The contribution to the Gibbs free energy has a strict mathematical form calculated from statistical mechanics [1]

$$g_{\text{ENTROPY}} = RT[x_A \ln(x_A) + x_B \ln(x_B)] \tag{3.2}$$

where $R = 8.314$ J mol^{-1} K^{-1} is the gas constant, and the temperature T is in Kelvin (K). This function is plotted as the red line in figure 3.1 using $x_A = 1 - x_B$ from

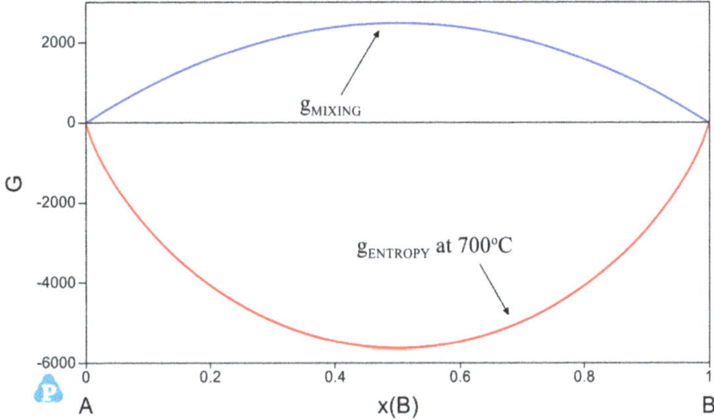

Figure 3.1. Entropic and mixing energies a function of the mole fraction of component B. Created using PANDAT [6].

equation (3.1). The entropic contribution is always negative and scales linearly with temperature. It has a minima at $x_A = x_B = 0.5$, i.e. a fully mixed solution of A and B. A system wishes to minimise its energy and hence entropy always promotes mixing. As the temperature increases, the entropy becomes more dominant over other contributions forcing more inter-mixing between the components. For a multi-component system, equation (3.2) can be extended to

$$g_{ENTROPY} = RT \sum_{j=1}^{M} x_i \ln(x_i).$$

3.1.2 Energy of mixing

A major contribution to the enthalpy of a system is from the potential energy due to bonding between atoms. For this idealised system the potential energy arises from A–A bonds, B–B bonds and A–B bonds. Each interatomic potential is different, and the values can either promote mixing (A–B bonds are favoured) or suppress it (A–B bonds are not favoured). At the continuum level this mixing energy is represented in its simplest form by a quadratic function

$$g_{MIXING} = Lx_A x_B. \tag{3.3}$$

This is plotted as the blue line in figure 3.1 for $L = 10$ kJ mol^{-1} using $x_A = 1 - x_B$ from equation (3.1). If the A–B bond energy is less than the A–A or B–B bond energies then there is a preference for A–B bonds to form. In this case mixing of the elements is promoted and $L < 0$. This reinforces the entropic contribution of equation (3.2) which is also driving mixing. Conversely, if the A–B bond has a higher energy than the A–A and B–B bonds then A and B will not wish to bond and $L > 0$. In this case mixing is suppressed but, as we shall see in section 3.1.6, entropy always ensures some mixing, even if it is just a tiny amount.

3.1.3 Total Gibbs free energy

The total Gibbs free energy is the sum of the entropic and mixing contributions of equations (3.2) and (3.3). These are often referred to as the **ideal energy** and **excess energy** respectively. This gives the **regular solution** model for the Gibbs free energy

$$g_S(x_i, T) = L_S x_A x_B + RT[x_A \ln x_A + x_B \ln (x_B)] \tag{3.4}$$

where the S subscript denotes this is the energy of the SOLID phase. When $L_S = 0$ this is known as the **ideal solution** model. Figure 3.2(a) shows the total Gibbs free

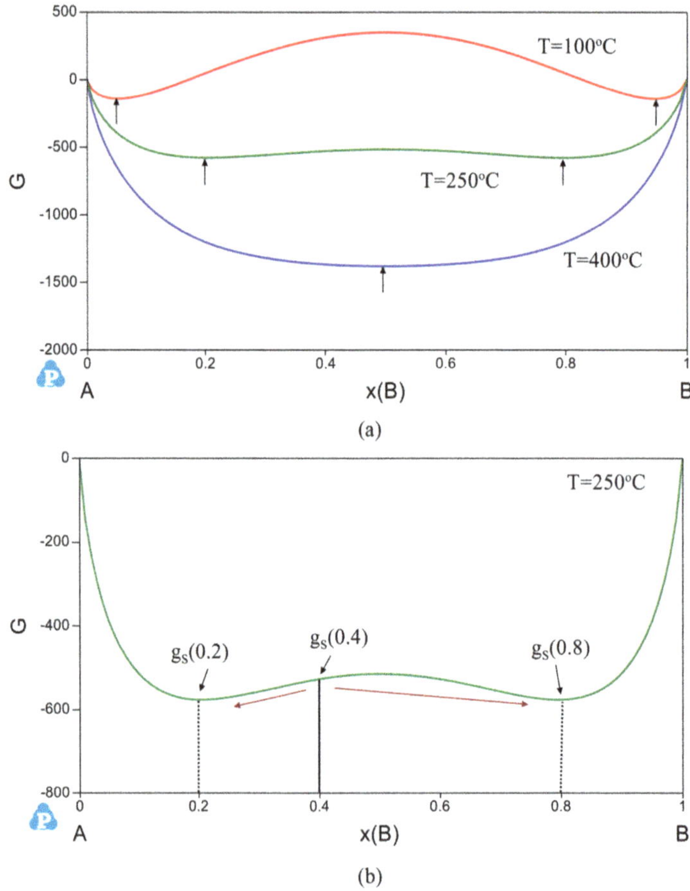

Figure 3.2. (a) Entropy dominates the Gibbs free energy at high temperatures (over 330 °C). Excess (mixing) energy changes the shape of the curve at lower temperatures (below 330 °C). Arrows indicate minimum energy states. (b) Illustration of phase separation at 250 °C. A two phase region exists where the nominal composition is between the minima at $x_B = 0.2$ and $x_B = 0.8$. For a nominal composition of $x_B = 0.4$, the red arrows show that the total energy can be reduced by forming two phases with the minimum energy compositions. Created using PANDAT [6].

energy at three different temperatures for a value of $L_S = 10$ kJ mol^{-1}. Mixing is promoted by the entropic term. This dominates at the highest temperature of 400 °C. At lower temperatures the excess energy becomes more important. The arrows in figure 3.2(a) indicate the location of the minimum energy states.

Consider an alloy with an overall **nominal composition** of B of $x_0 = 0.4$. At a given temperature, the system will adapt itself to minimise the total Gibbs free energy. In some cases a system can undergo a **phase separation** to reduce this energy, whereby two phases exist simultaneously. In this single phase example, one of these phases would have a higher (richer) concentration of B ($x_B > 0.4$) and one would have a lower (poorer) concentration of B ($x_B < 0.4$). At 400 °C there is just one energy minima so a phase separation is not energetically favourable, and the system will exist as a single solution with composition $x_B = 0.4$. Two minima appear in the energy profile as the excess energy becomes more influential at lower temperatures. At 250 °C these minima are at compositions of 0.2 and 0.8. If the nominal composition of the alloy is between these two minima (i.e. $0.2 < x_0 < 0.8$) then there will be a phase separation, forming a phase with $x_B = 0.2$ and a phase with $x_B = 0.8$. This is illustrated in figure 3.2(b), as $g_S(0.2) = g_S(0.8) < g_S(0.4)$. If the nominal composition is outside these minima (i.e. $x_0 < 0.2$ or $x_0 > 0.8$) then phase separation is not favourable as the overall energy cannot be reduced by this process. This observation is derived mathematically in section 3.2.2.

3.1.4 A simple TDB file

The thermodynamics of the system are defined as a function of temperature and composition by equation (3.4). The lowest energy state can be found at each temperature, and the regions where different phases exist plotted in a temperature versus composition plot known as a **phase diagram**. A number of commercial programs, such as PANDAT or Thermocalc, can do this, as described in section 1.4.1. The energy of each phase is defined by a Thermodynamics DataBase file, denoted by the extension TDB. The TDB file has a fixed notation, the basics of which are introduced here.

The general Gibbs free energy model for a regular solution of phase p has some extra terms to (3.4) which must be defined in the TDB file

$$g_p(x_i, T) = g_p^A x_A + g_p^B x_B + L_p x_A x_B + RT(x_A \ln(x_A) + x_B \ln(x_B)) \tag{3.5}$$

where $g_p^A(T)$ and $g_p^B(T)$ are the temperature dependent energies in phase p of pure A and pure B respectively, and the excess energy parameter is extended to a higher order polynomial fit using the Redlich–Kister formulation

$$L_p = L_p^{(0)} + L_p^{(1)}(x_A - x_B) + L_p^{(2)}(x_A - x_B)^2 + L_p^{(3)}(x_A - x_B)^3 \tag{3.6}$$

where the coefficients $L_p^i(T)$ can also be functions of temperature. The inclusion of these higher order terms indicates that the potential energy of an atom will depend on complex interactions with all the atoms in its immediate environment, not just the simple pair-wise interactions considered in the formulation of equation (3.3). These terms are defined in the following file called `onephase.tdb`.

```
Thermodynamic database file #1: onephase.tdb

ELEMENT A      SOLID     1.0 0.0 0.0 !
ELEMENT B      SOLID     1.0 0.0 0.0 !
PHASE SOLID % 1 1 !
  CONSTITUENT SOLID : A B : !
  PARAMETER G(SOLID,A) 298.15 0.0; 3000.0 N !
  PARAMETER G(SOLID,B) 298.15 0.0; 3000.0 N !
  PARAMETER L(SOLID,A,B;0) 298.15 10000.0; 3000.0 N !
```

The **ELEMENT** command defines the element name (A or B in this case), its most stable (reference) phase (there is only one phase called SOLID here), its mass (assumed to be 1.0) and its reference enthalpy and entropy (both taken to be zero here). All real elements have standard definitions, which are discussed in more detail in section 3.4. Each finished line must end in an exclamation mark (!).

The **PHASE** command defines each phase. There is just one phase in this case. Any name that is not a keyword is allowed. Here the phase is simply called SOLID. The numbers after the % sign define the crystal model for this phase. The first one indicates that there is one sub-lattice and the second one that there is one site on this sub-lattice for atoms to occupy. Sub-lattice models are discussed in more detail in section 3.3.

The **CONSTITUENT** command defines which elements can occupy the first (and only) sub-lattice in the phase SOLID, namely A and B.

The **PARAMETER** command defines temperature dependent parameters in the thermodynamic model of equation (3.5). Different functions can be provided for different temperature ranges, with the lower temperature before the function and the upper temperature after it. Here the standard temperature ranges of 298.15 K (25 °C) to 3000 K are used. An N is used to end the function definition, and a Y is used if a further line is to be added to define the function over another temperature range. Denoting the SOLID phase by the symbol S, the above definitions state that $g_S^A(T) = 0$ in the first parameter line, $g_S^B(T) = 0$ in the second line and the zeroth Redlich–Lister parameter $L_S^{(0)} = 10$ kJ mol^{-1} in the third line. These key numbers are highlighted in bold.

This code therefore defines the regular solution model of equation (3.4). Note that the last (entropic) term in (3.4) and (3.5) does not need to be defined explicitly, as this is the same for all phases and is included by default.

3.1.5 A simple binary phase diagram

The Gibbs free energy defined by equation (3.4) defines the energy profile at different temperatures, as shown in figure 3.2(a). In this case, due to the symmetry of the profiles, the location of the energy minima demarcate the phase boundaries and can be plotted using PANDAT as a continuous function of temperature to produce a phase diagram.

PANDAT: plotting a binary phase diagram using 'onephase.tdb'
1. Copy and paste the contents of 'onephase.tdb' from section 3.1.4 into a text editor and save it into a file of that name.
2. Run Pandat. Under 'Create a New Workspace', select the 'PanPhaseDiagram' icon and click the 'Create' button.
3. Select 'Databases → Load TDB or PDB (Encrypted TDB)' and open the file 'onephase.tdb'.
4. Under 'Select Components', click on the 'Sel/Clr All' button to move all the 'Available Components' into the 'Selected Components' column. Press 'OK'.
5. Select 'PanPhaseDiagram → Section Calculation'.
6. The resulting screen allows you to change the axes of the phase diagram by setting the temperature at three points on the axes. The default values are generally acceptable so press 'OK' to plot the phase diagram.
7. Once the diagram has appeared, select Graph → Label and click in the different regions to label them. Note that in this example the two regions both have the same name 'SOLID'. If you wish, you can also augment the diagram by editing the labels, drawing lines and adding text as shown in figure 3.3.

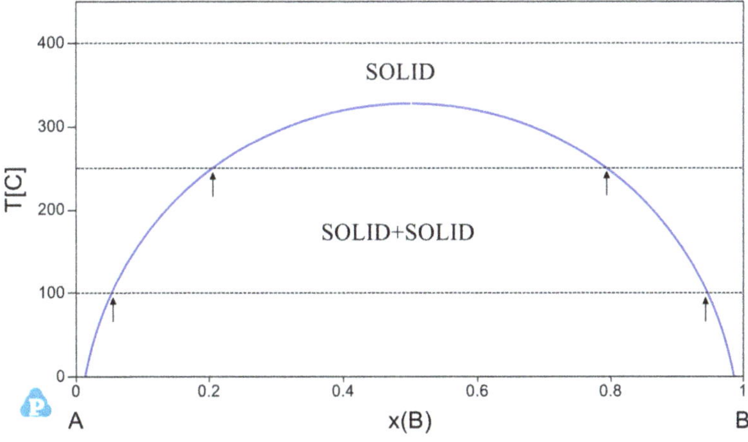

Figure 3.3. A simple phase diagram for a one phase (SOLID) and two component (A and B) binary system. Created using PANDAT [6].

The resulting phase diagram is shown in figure 3.3. The locations of the minima on the two isotherms at 250 °C and 100 °C are transferred from figure 3.2(a) and show that they correspond to boundaries between the two different phase regions. The phase boundary does not intersect the isotherm at 400 °C as there is only one energy minima at this temperature. At first glance it may appear strange that two phase regions can appear when there is strictly only one phase. Here the distinction between phases is related to their composition only. The uppermost SOLID region consists of a single solid solution phase of uniform composition where A and B are fully mixed. The SOLID + SOLID region consists of two different solid solutions, where the composition of B in each is given by the locations of the two minima in the Gibbs free energy plot. As described in section 3.1.3, at 250 °C the minima are at $x_B = 0.2$ and $x_B = 0.8$. This means that, if the nominal composition is between 20% and 80% then the material will consist of two solid phases: one phase is 20% B and 80% A, and the other is 80% B and 20% A. The amount of each phase is determined by the **lever rule**, which is introduced in section 3.2.3.

3.1.6 Chemical potential

The chemical potential, μ_i^p, represents how a change in the concentration of a component i changes the energy of a given phase p. It is defined by writing the Gibbs free energy of the phase as a linear function of its composition

$$g_p(x_A, x_B, T) = \mu_A^p x_A + \mu_B^p x_B \tag{3.7}$$

where constant pressure and temperature are assumed. Figure 3.4 shows that the two chemical potentials, μ_A^p and μ_B^p can be found from where the tangent to the Gibbs free energy for phase p at a given temperature and composition intercepts the $x_A = 1$ (left) and $x_B - 1$ (right) axes respectively. The tangent, and hence the chemical potentials, change with the composition of the alloy. As we shall see in section 6.1, variations in the chemical potential of a substance are the driving force for microstructural change. Hence at equilibrium the chemical potential of a substance must be the same everywhere.

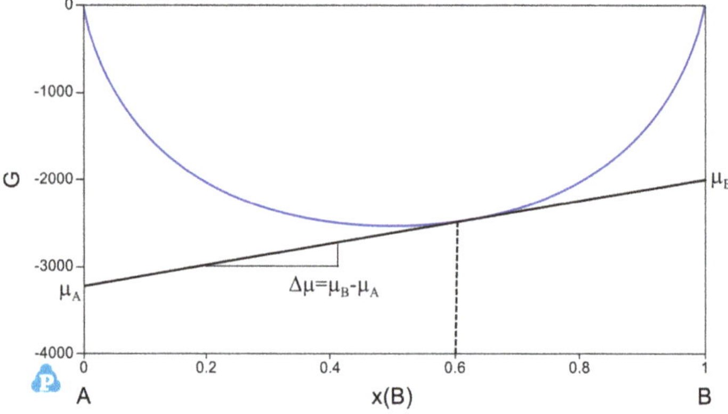

Figure 3.4. The chemical potentials of A and B at a composition $x_B = 0.6$ are shown. They are the intercepts of the tangent to the Gibbs free energy curve at this point. The slope of the tangent is $\Delta\mu = \mu_B - \mu_A$. Created using PANDAT [6].

The slope of the Gibbs free energy curve at x_B is the chemical potential difference

$$\left. \frac{\partial g_p}{\partial x_B} \right|_{x_A = 1 - x_B} = \mu_B^p - \mu_A^p = \Delta\mu^p. \qquad (3.8)$$

Hence the chemical potentials are

$$\mu_A^p = g_p - x_B \frac{\partial g_p}{\partial x_B} \qquad \mu_B^p = g_p + x_A \frac{\partial g_p}{\partial x_B}.$$

For the regular solution model of equation (3.4) with constant L_p this yields

$$\mu_A^p = g_p^A + L_p(1 - x_A)^2 + RT \ln(x_A)$$
$$\mu_B^p = g_p^B + L_p(1 - x_B)^2 + RT \ln(x_B). \qquad (3.9)$$

Now reconsider figure 3.1 for the ideal (entropic) and excess (mixing) energies. Superficially the curves look a similar shape, although bowing in a different direction. This is not the case however. The difference is very clear if we plot the contributions to the chemical potential from each.

From equation (3.2), the contribution to the chemical potential difference due to entropy is

$$\Delta\mu_{\text{ENTROPY}} = \left. \frac{\partial g_{\text{ENTROPY}}}{\partial x_B} \right|_{x_A = 1 - x_B} = RT \ln\left(\frac{x_B}{1 - x_B} \right)$$

and similarly, from equation (3.3), the contribution from mixing is

$$\Delta\mu_{\text{MIXING}} = \left. \frac{\partial g_{\text{MIXING}}}{\partial x_B} \right|_{x_A = 1 - x_B} = L(1 - 2x_B).$$

These two contributions to the chemical potential are both plotted in figure 3.5 for $L = -20$ kJ mol^{-1} and $T = 700\,^{\circ}$C. This value of L has been chosen as the two Gibbs free energy curves look almost identical when plotted. However, when plotting the chemical potential difference (slopes) the behaviour at very low and very high compositions is clearly very different. The entropic chemical potential goes to minus infinity at $x_B = 0$ and plus infinity at $x_B = 1$. The contribution from mixing always remains finite.

This demonstrates the fact that entropy will always win against the excess energy contribution at these two extremes, i.e. the entropic contribution to the chemical potential that forces components to mix can never be completely stopped (although the amount of mixing can become very small, particularly at low temperatures). This also shows that there is an infinite thermodynamic barrier at the extreme values of the compositional range, such that a system can never evolve such that its composition goes below 0 or above 1. Intuitively we know this must always be true, but mathematically such things cannot be taken for granted, so it is reassuring that the entropy makes this a physical condition.

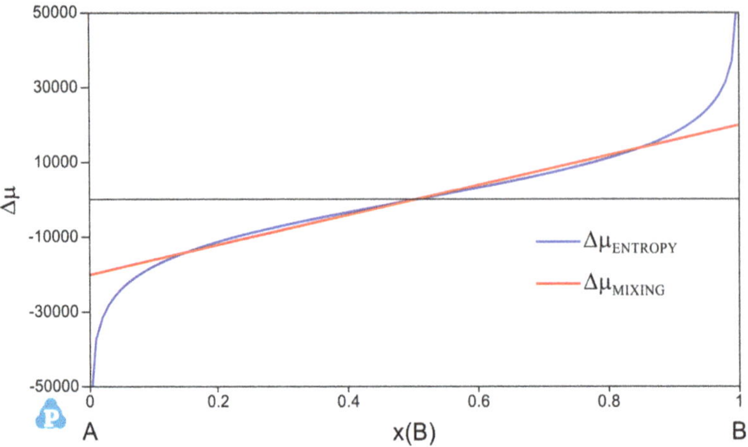

Figure 3.5. Entropic and mixing energy contributions to chemical potential difference. Created using PANDAT [6].

For a general M component system the Gibbs free energy of a phase p can be expressed as

$$g_{\text{p}}(x_i, T) = \sum_{i=1}^{M} \mu_i^{\,p} x_i.$$

3.2 A two phase system with two components

A second phase called LIQUID is now introduced, in addition to the SOLID phase. They are both formed from the two elements A and B. The inclusion of a second phase allows for a much greater variety in the topology of the phase diagrams. The Gibbs free energy of the SOLID phase is defined by equation (3.5) using equation (3.1) such that

$$g_{\text{S}} = L_{\text{S}}(1 - x_{\text{B}})x_{\text{B}} + RT((1 - x_{\text{B}}) \ln (1 - x_{\text{B}}) + x_{\text{B}} \ln (x_{\text{B}})). \tag{3.10}$$

A similar expression is used to define the Gibbs free energy of the LIQUID phase (with respect to that of the SOLID phase) such that

$$\begin{aligned}
g_{\text{L}} = {} &g_{\text{A}}^{\text{L}}(1 - x_{\text{B}}) + g_{\text{B}}^{\text{L}}x_{\text{B}} + L_{\text{L}}(1 - x_{\text{B}})x_{\text{B}} \\
&+ RT((1 - x_{\text{B}}) \ln (1 - x_{\text{B}}) + x_{\text{B}} \ln (x_{\text{B}})).
\end{aligned} \tag{3.11}$$

If the melting temperature of pure A is T_{A}^{M} and that of pure B is T_{B}^{M} then equation (2.6) can be adapted to write the two temperature dependent parameters that define the two melting points as

$$g_{\text{A}}^{\text{L}} = \Delta S_{\text{A}}(T_{\text{A}}^{\text{M}} - T) \qquad g_{\text{B}}^{\text{L}} = \Delta S_{\text{B}}(T_{\text{B}}^{\text{M}} - T). \tag{3.12}$$

For the purposes of these calculations we choose

$$T_A^M = 800\ {}^\circ\text{C} \qquad T_B^M = 1000\ {}^\circ\text{C} \qquad \Delta S_A = \Delta S_B = 20\ \text{J mol}^{-1}\ \text{K}^{-1}$$

and explore how the phase diagram can be affected by changing the mixing energies in the liquid and the solid, defined by L_S and L_L.

The additional LIQUID phase is added to the TDB file of section 3.1.4 to create a new file `twophase.tdb`. The LIQUID phase definition is followed by ':L', to denote that it is a liquid phase. Similarly, a ':G' is used after a gas phase. The important parameters are shown in bold. The code is shown for the case of $L_S = 25$ kJ mol^{-1} and $L_L = -20$ kJ mol^{-1} (both underlined). The rest of this section explores how these two parameters affect the topology of a binary phase diagram.

Thermodynamic database file #2: twophase.tdb

```
ELEMENT A     SOLID                    1.0 0.0 0.0 !
ELEMENT B     SOLID                    1.0 0.0 0.0 !

PHASE SOLID % 1 1 !
  CONSTITUENT SOLID : A B : !
  PARAMETER G(SOLID,A) 298.15 0.0; 3000.0 N !
  PARAMETER G(SOLID,B) 298.15 0.0; 3000.0 N !
  PARAMETER L(SOLID,A,B;0) 298.15 25000.0; 3000.0 N !

PHASE LIQUID:L % 1 1 !
  CONSTITUENT LIQUID:L : A B : !
  PARAMETER G(LIQUID,A) 298.15 20.0*(800.0+273.15-T); 3000.0 N !
  PARAMETER G(LIQUID,B) 298.15 20.0*(1000.0+273.15-T); 3000.0 N !
  PARAMETER L(LIQUID,A,B;0) 298.15 -20000.0; 3000.0 N !
```

3.2.1 Phase diagram for case with zero excess energy

This is the simplest case of $L_S = L_L = 0$. The binary phase diagram is plotted as follows.

PANDAT: plotting a binary phase diagram using 'twophase.tdb'
1. Copy and paste the contents of 'twophase.tdb' from section 3.2 into a text editor.
2. Edit the file by changing L_s from 25000.0 to 0.0 in the SOLID phase definition of parameter L.
3. Edit the file by changing L_L from −20000.0 to 0.0 in the LIQUID phase definition of parameter L.
4. Save it into a file called 'twophase.tdb'.
5. Run Pandat. Under 'Create a New Workspace', select the 'PanPhaseDiagram' icon and click the 'Create' button.

6. Select 'Databases → Load TDB or PDB (Encrypted TDB)' and open the file 'twophase.tdb'.
7. Under 'Select Components', click on the 'Sel/Clr All' button to move all the 'Available Components' into the 'Selected Components' column. Press 'OK'.
8. Select 'PanPhaseDiagram → Section Calculation'. Press 'OK'.
9. Use 'Graph-Label' to label each region.

The phase diagram is shown in figure 3.6. There is a LIQUID region at high temperatures, below that there is a transitional LIQUID + SOLID region where the alloy is in the 'mushy' zone of being part liquid and part solid, and a fully SOLID region at temperatures below that. Note that the width of the LIQUID + SOLID region tapers to zero at its left-hand and right-hand ends. This is because, as introduced for aluminium in chapter 2, a pure metal (pure A or pure B in this case) exhibits a sharp melting transition with a well-defined melting temperature. As specified in the TDB file, the melting temperatures for pure A and B are 800 °C and 1000 °C respectively. For alloys the entropy spreads the melting transition over a temperature range, where the fraction of liquid in the two phase liquid-solid melt increases with the temperature. Isotherms at 100 °C intervals (dashed lines) have been added to the plot. The Gibbs free energy variation along these isotherms is plotted in figure 3.7 (using a 'Line Calculation') to demonstrate how the phase diagram is constructed.

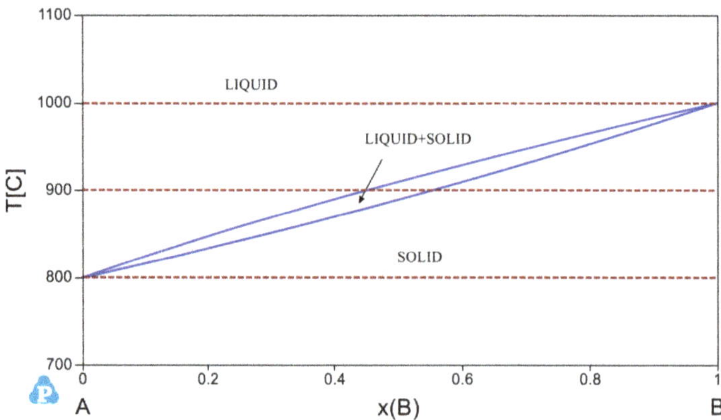

Figure 3.6. A basic two phase binary phase diagram with zero (excess) mixing energy ($L_S = L_L = 0$). The Gibbs free energy variation along the isotherms (dashed lines) are shown in figure 3.7. Created using PANDAT [6].

PANDAT: plotting phase energies for constant temperature

1. Select 'PanPhaseDiagram → Line Calculation'.
2. Select 'Options'. Uncheck any checked boxes in the 'Choose *Y*-axis Properties' column and check the 'G(@*)' checkbox. (If it is not already there, check it in the 'Choose table Columns' column also). Press 'OK'.
3. Under 'Start Point' change the temperature 'T(C)' to '700', the composition of A 'x(A)' to 1 and the composition of B 'x(B)' to 0. Note that the 'Total' composition must always add up to 1.
4. Under 'End Point' change the temperature 'T(C)' to '700', the composition of A 'x(A)' to 0 and the composition of B 'x(B)' to 1.
5. Make sure that the 'Individual Phases' checkbox is checked. Press 'OK'.

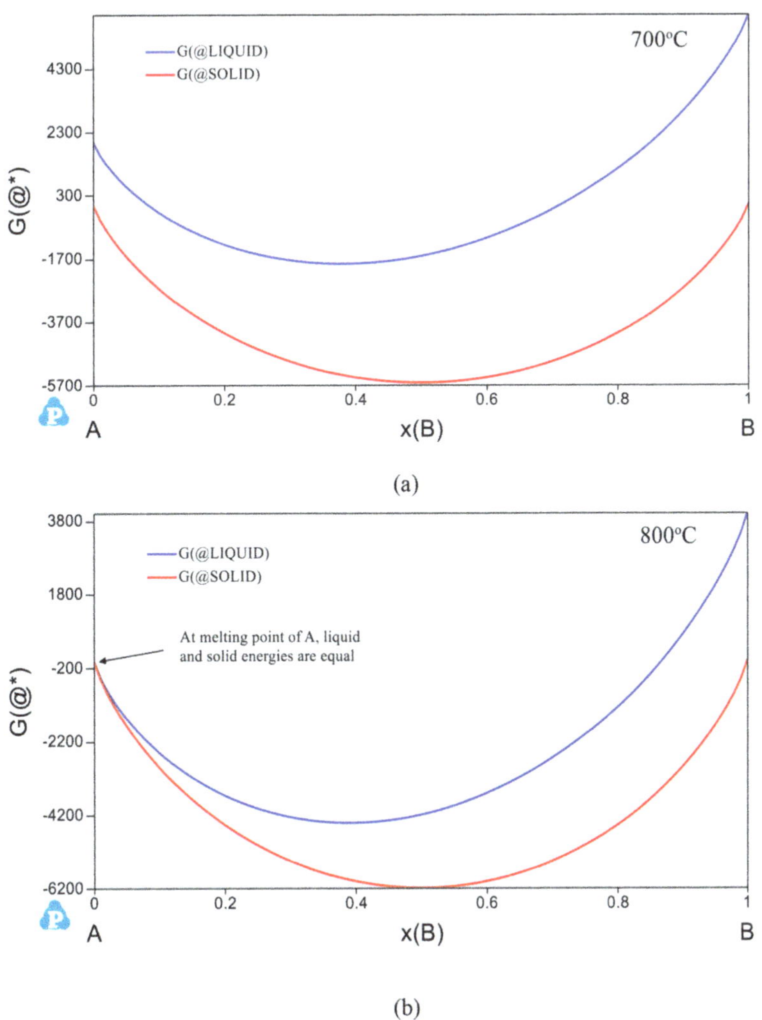

Figure 3.7. Gibbs free energy profiles along the isotherms shown in figure 3.6 at (a) 700 °C, (b) 800 °C, (c) 900 °C, (d) 1000 °C, (e) 1100 °C. Created using PANDAT [6].

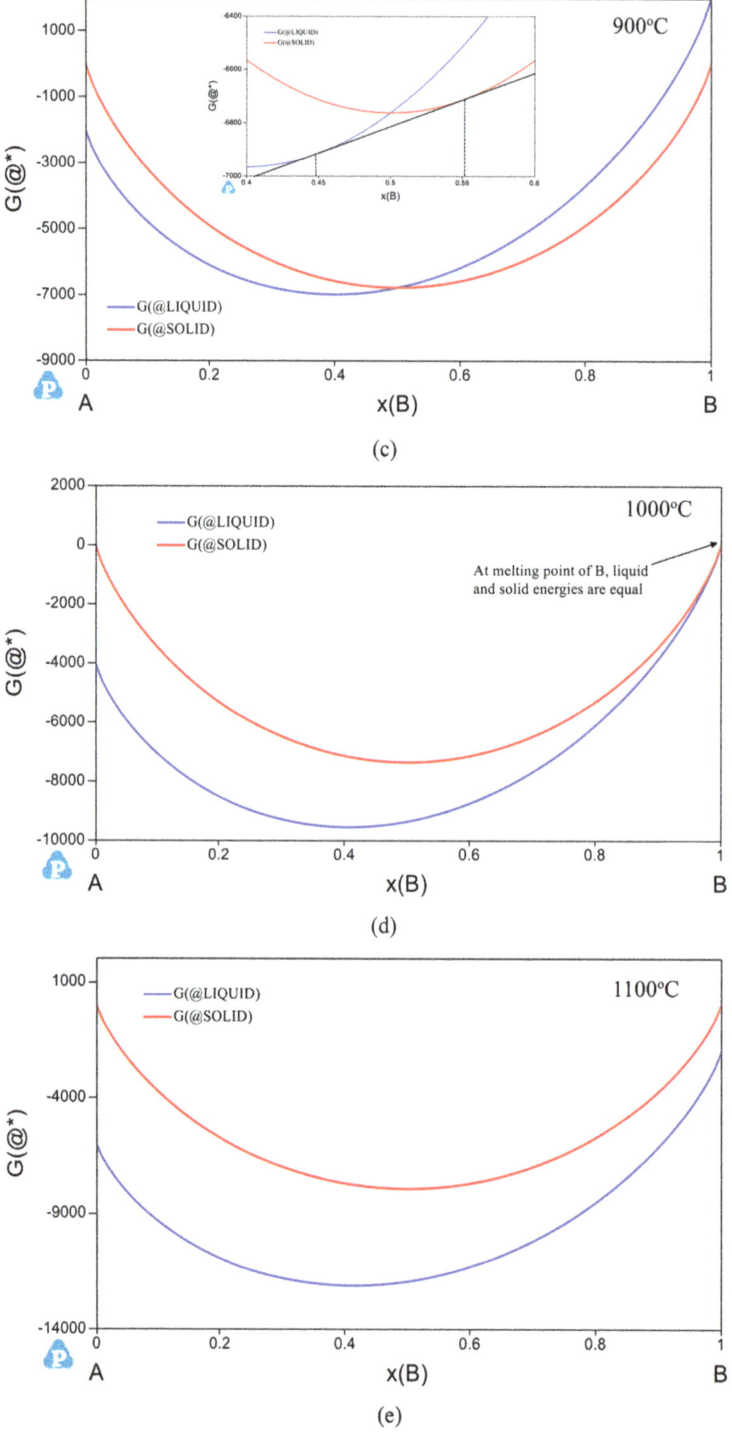

Figure 3.7. (Continued.)

At 700 °C the energy of the SOLID phase is always below that of the LIQUID phase and therefore the SOLID phase exists at all compositions at this temperature, as shown in figure 3.6. At 800 °C, the melting temperature of pure A, the LIQUID and SOLID curves touch at $x_A = 1$. Any further increase in temperature will push the LIQUID curve below the SOLID curve at $x_A = 1$ signifying the transition from SOLID to LIQUID of pure A at this temperature. What happens at 900 °C is not so obvious as the LIQUID phase has the lowest energy at some points, and the SOLID phase has the lowest energy at others. This also coincides with the appearance of the two phase region in the phase diagram (LIQUID + SOLID). The inset to figure 3.7(c) shows an enlargement of the two phase region. The common tangent line for the two curves is drawn. The phase boundaries are where the line touches the two curves, at roughly $x_B = 0.44$ and $x_B = 0.56$. The general procedure for finding the minimum energy in such two phase regions is derived in the next section 3.2.2. The situation at 1000 °C is similar to that at 800 °C, except now the energy curves meet at $x_B = 1$ as this is the melting temperature for pure B. At 1100 °C the LIQUID energy curve is always below the SOLID energy curve, indicating that the alloy is LIQUID at all compositions at this temperature.

3.2.2 Equilibrium conditions in two phase regions

Two phase regions in a binary phase diagram are where two phases exist at once. To calculate the phase boundaries of these regions in our model system, let x_B^L be the mole fraction of B in the LIQUID phase and x_B^S be the mole fraction of B in the SOLID phase. Let f_S be the mole fraction of the SOLID phase, and f_L be the mole fraction of the LIQUID phase. For a given temperature, the Gibbs free energies per mole for each phase are $g_L(x_B^L)$ and $g_S(x_B^S)$. The total Gibbs free energy per mole of the two phase mixture is then

$$g_{TOT} = f_L g_L(x_B^L) + f_S g_S(x_B^S). \tag{3.13}$$

At equilibrium the Gibbs free energy is minimised, subject to the total amount of A and B remaining constant. Let x_0 be the nominal mole fraction of B in the material of interest. Then we must have

$$f_L x_B^L + f_S x_B^S = x_0. \tag{3.14}$$

Given that $f_S + f_L = 1$, there are three unknowns (f_S, x_B^L and x_B^S) to determine.

Derivation

At equilibrium, the state does not change over time. Hence $\dfrac{dg_{TOT}}{dt} = 0$ or, to use the dot notation, $\dot{g}_{TOT} = 0$.

Using the chain rule for differentiation, (3.13) gives

$$\dot{g}_{TOT} = -\dot{f}_S g_L + (1 - f_S)\frac{dg_L}{dx_B^L}\dot{x}_B^L + \dot{f}_S g_S + f_S \frac{dg_S}{dx_B^S}\dot{x}_B^S = 0. \tag{3.15}$$

Using the definition for the chemical potential difference (3.8) we therefore have the condition

$$\dot{f}_{\mathrm{S}}(g_{\mathrm{S}} - g_{\mathrm{L}}) + (1 - f_{\mathrm{S}})\Delta\mu^{\mathrm{L}}\dot{x}_{\mathrm{B}}^{\mathrm{L}} + f_{\mathrm{S}}\Delta\mu^{\mathrm{S}}\dot{x}_{\mathrm{B}}^{\mathrm{S}} = 0.$$

This is subject to the condition (3.14) which in rate form is

$$-\dot{f}_{\mathrm{S}}x_{\mathrm{B}}^{\mathrm{L}} + (1 - f_{\mathrm{S}})\dot{x}_{\mathrm{B}}^{\mathrm{L}} + \dot{f}_{\mathrm{S}}x_{\mathrm{B}}^{\mathrm{S}} + f_{\mathrm{s}}\dot{x}_{\mathrm{B}}^{\mathrm{S}} = 0$$

or

$$(1 - f_{\mathrm{S}})\dot{x}_{\mathrm{B}}^{\mathrm{L}} = \dot{f}_{\mathrm{S}}(x_{\mathrm{B}}^{\mathrm{L}} - x_{\mathrm{B}}^{\mathrm{S}}) - f_{\mathrm{S}}\dot{x}_{\mathrm{B}}^{\mathrm{S}}.$$

Substituting this constraint into the minimum energy condition of equation (3.15) yields

$$[g_{\mathrm{S}} - g_{\mathrm{L}} + \Delta\mu^{\mathrm{L}}(x_{\mathrm{B}}^{\mathrm{L}} - x_{\mathrm{B}}^{\mathrm{S}})]\dot{f} + [\Delta\mu^{\mathrm{S}} - \Delta\mu^{\mathrm{L}}]f\dot{x}_{\mathrm{B}}^{\mathrm{S}} = 0. \tag{3.16}$$

Equation (3.16) is always satisfied when both the expressions in the square brackets are zero such that

$$\Delta\mu^{\mathrm{L}} = \Delta\mu^{\mathrm{S}} = \frac{g_{\mathrm{L}}(x_{\mathrm{B}}^{\mathrm{L}}) - g_{\mathrm{S}}(x_{\mathrm{B}}^{\mathrm{S}})}{x_{\mathrm{B}}^{\mathrm{L}} - x_{\mathrm{B}}^{\mathrm{S}}}. \tag{3.17}$$

These are the two conditions required for equilibrium in a two phase region. It should be noted that (3.14) only applies when the nominal composition sits between the equilibrium compositions in the two phases, i.e. when $x_{\mathrm{B}}^{\mathrm{L}} \leqslant x_0 \leqslant x_{\mathrm{B}}^{\mathrm{S}}$ or $x_{\mathrm{B}}^{\mathrm{S}} \leqslant x_0 \leqslant x_{\mathrm{B}}^{\mathrm{L}}$.

The conditions specified by (3.17) can be satisfied graphically, as shown in figure 3.8. The equilibrium compositions in the solid and liquid, $x_{\mathrm{B}}^{\mathrm{S}}$ and $x_{\mathrm{B}}^{\mathrm{L}}$, occur at the points where the common tangent line touches the solid and liquid energy curves. The definition of the chemical potential, illustrated in figure 3.4, clearly shows that the equilibrium condition (3.17) can also be written as

$$\mu_{\mathrm{A}}^{\mathrm{L}} = \mu_{\mathrm{A}}^{\mathrm{S}} \qquad \mu_{\mathrm{B}}^{\mathrm{L}} = \mu_{\mathrm{B}}^{\mathrm{S}} \tag{3.18}$$

i.e. the chemical potential of each component is the same in all phases at equilibrium.

3.2.3 The lever rule

Equation (3.14) can be rearranged to show that the ratio of the distances between the composition lines gives the solid fraction

$$f_{\mathrm{S}} = \frac{x_0 - x_{\mathrm{B}}^{\mathrm{L}}}{x_{\mathrm{B}}^{\mathrm{S}} - x_{\mathrm{B}}^{\mathrm{L}}} \tag{3.19}$$

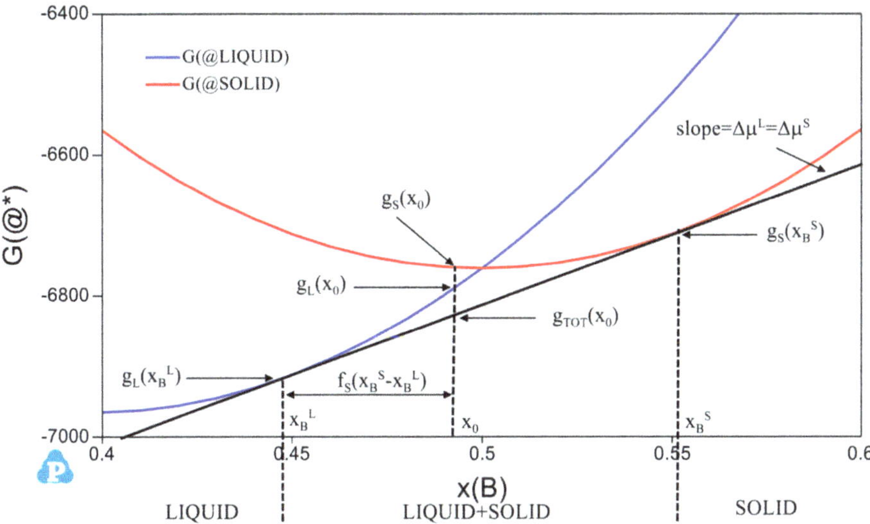

Figure 3.8. Graphical determination of the equilibrium condition for the two phase region between x_B^L and x_B^S. Created using PANDAT [6].

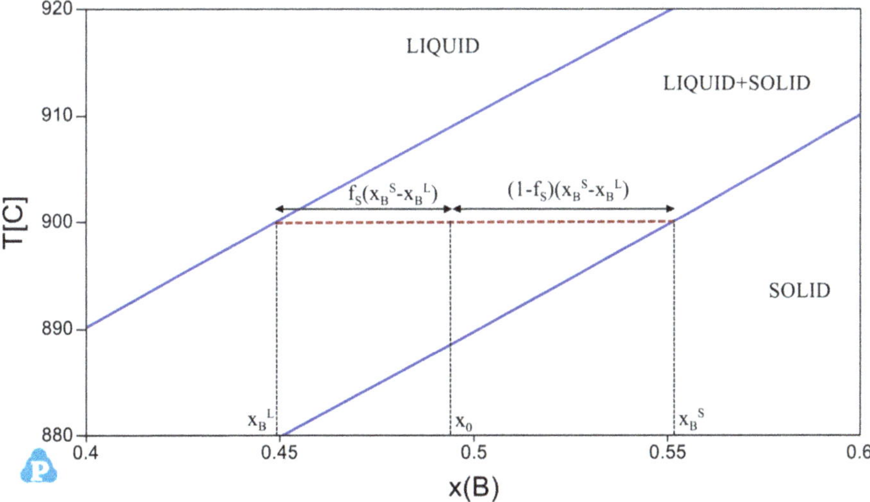

Figure 3.9. The phase diagram of figure 3.6, magnified around the 900 °C isotherm phase boundaries, illustrating the use of levers rule to calculate the molar fraction f_S of the solid phase on the (red dashed) tie line. Created using PANDAT [6].

and therefore also the liquid fraction as $f_L = 1 - f_S$. This is known as **levers rule**. It allows the molar fraction of any phase in a two phase region to be graphically measured from the phase diagram. This process is illustrated in figure 3.9 for the case in figure 3.7. The horizontal red dashed line is known as a **tie line**, which is a line in a

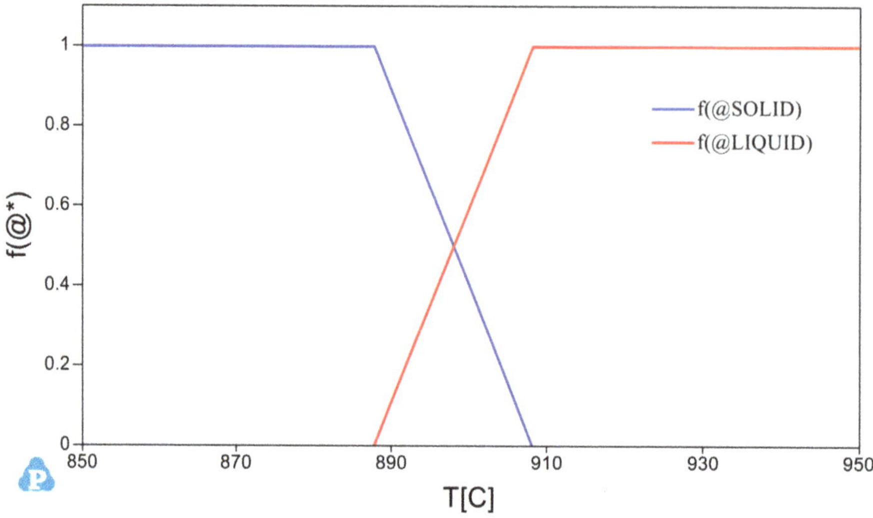

Figure 3.10. The molar fractions of the SOLID and LIQUID phases for a nominal composition of 0.49 for B. The melting phase transition occurs over a 20 °C range, from 898 °C to 908 °C. Created using PANDAT [6].

two phase region of a phase diagram where levers rule can be used. PANDAT can also plot the phase fraction over a temperature range. The following chooses a nominal composition for B of $x_0 = 0.49$ and plots both phase fractions over a temperature range of 850 °C to 950 °C, as shown in figure 3.10.

> **PANDAT: plotting phase fraction over a temperature range**
> 1. Continuing the previous analysis, select 'PanPhaseDiagram → Line Calculation'.
> 2. Select 'Options'. Uncheck any checked boxes in the 'Choose Y-axis Properties' column and check the 'f(@*)' checkbox. (If it is not already there, check it in the 'Choose table Columns' column also.) Press 'OK'.
> 3. Under 'Start Point' change the temperature 'T(C)' to '850', the composition of A 'x(A)' 'to 0.51' and the composition of B 'x(B)' to '0.49'. Note that the 'Total' composition must always add up to 1.
> 4. Under 'End Point' change the temperature 'T(C)' to '950', the composition of A 'x(A)' to '0.51' and the composition of B 'x(B)' to '0.49'.
> 5. Make sure that the 'Individual Phases' checkbox is NOT checked. Press 'OK'.

Equation (3.13) and levers rule show that the total Gibbs free energy of the system, g_{TOT}, is the point on the tangent line where the composition of B has the nominal value, x_0. This is illustrated in figure 3.8, and clearly shows that the separation into two phases lowers the total energy below that of the individual phases, i.e. $g_{TOT}(x_0) < g_S(x_0)$ and $g_{TOT}(x_0) < g_L(x_0)$.

3.2.4 Effect of excess (mixing) energies on phase diagrams (exercise)

Here we explore how simply changing the excess energy coefficients in the liquid (L_L) and solid (L_S) phases can change the shape and topology of a simple binary phase diagram. It may be surprising how just changing these two parameters in a two phase model can generate the broad range of topologies exhibited by most binary systems. It is recommended that the reader try plotting the following phase diagrams using an appropriately adapted 'twophase.tdb' input file before consulting the results in figure 3.11.

(a) $L_S = 10000$ J mol^{-1} and $L_L = 10000$ J mol^{-1}.
(b) $L_S = 20000$ J mol^{-1} and $L_L = 20000$ J mol^{-1}.
(c) $L_S = 20000$ J mol^{-1} and $L_L = 0$ J mol^{-1}.
(d) $L_S = 10000$ J mol^{-1} and $L_L = 25000$ J mol^{-1}.
(e) $L_S = 25000$ J mol^{-1} and $L_L = 25000$ J mol^{-1}.

PANDAT: plotting a set of phase diagrams
1. Following on from the previous session, change the values of L_s and L_L in the file `twophase.tdb` to those required.
2. Save the file `twophase.tdb`.
3. In PANDAT, select 'Databases → Refresh the Active TDB' to upload the new changes. (This assumes `twophase.tdb` was already loaded in the previous session.) Include all the components as active.
4. Use 'PanPhaseDiagram → Section Calculation' to plot the phase diagram as usual.

The series of phase diagrams are plotted in figure 3.11. Note that it is merely the aim of this book to teach the underlying thermodynamics of phase diagrams. The reader is therefore referred to some of the many excellent resources for teaching the interpretation of phase diagrams [2, 3].

Recall that a positive L parameter acts to separate components, whereas a negative parameter reinforces the entropic contribution and forces them to mix. Figure 3.11(a) is roughly a combination of figures 3.3 and 3.6, with a small but positive value of L_S generating a SOLID + SOLID phase separation at low temperatures. The basic features of this phase diagram are similar to those of the Cr–Fe system (which readers can download from the NIMS database [4], as explained in section 1.4.1).

The values of L_S and L_L are doubled for the plot in figure 3.11(b). This expands both the upper LIQUID + SOLID region and the lower SOLID + SOLID region so that they impinge on one another. A horizontal boundary appears at this interface. This is a common occurrence in phase diagrams, and the reason for this is explained in the next phase diagram.

Figure 3.11(c) demonstrates some of the most commonly observed features of a binary phase diagram, where the energetics do not dissuade mixing in the liquid but actively discourage mixing in the solid. This is a known as a **eutectic phase diagram**, as it demonstrates a **eutectic point** (denoted by the letter E). A real example is the Pb–Sn phase diagram shown in figure 3.17. This is where three phases co-exist at a given

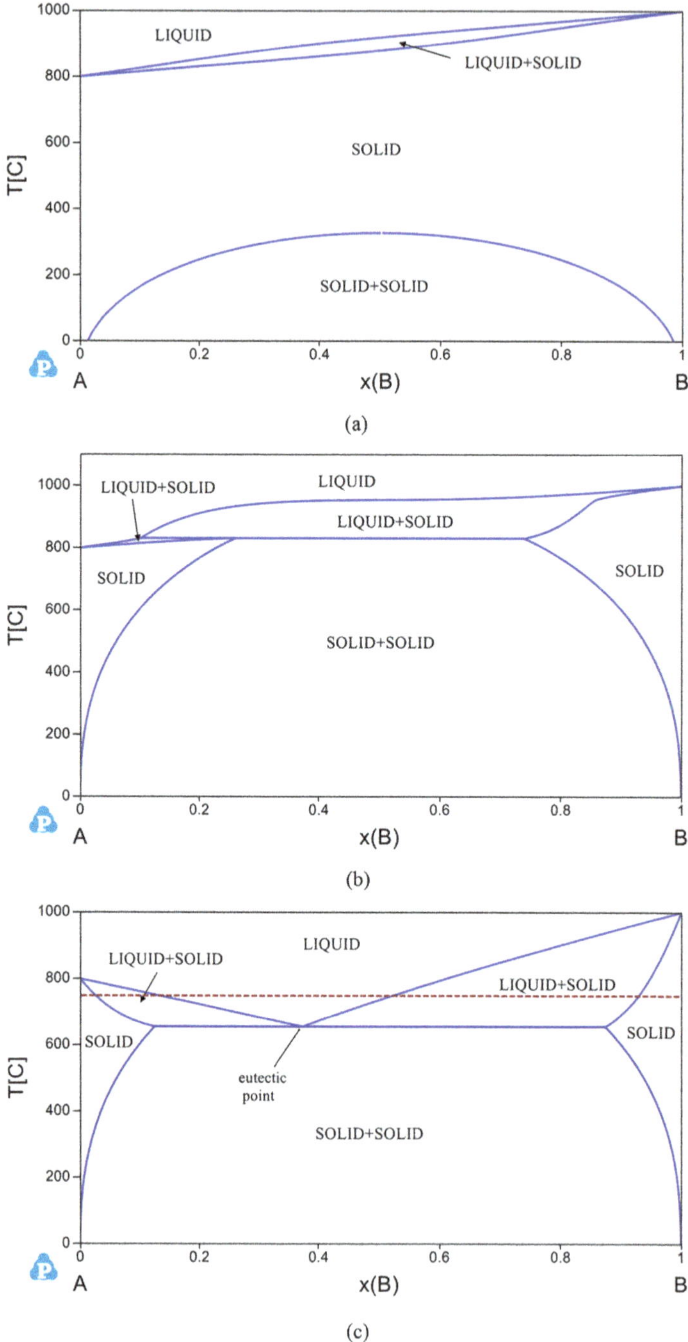

Figure 3.11. Binary phase diagrams produced from the file `twophase.tdb` with (a) $L_S = 10$ kJ mol^{-1} and $L_L = 10$ kJ mol^{-1}, (b) $L_S = 20$ kJ mol^{-1} and $L_L = 20$ kJ mol^{-1}, (c) $L_S = 20$ kJ mol^{-1} and $L_L = 0$ kJ mol^{-1}. (d) $L_S = 10$ kJ mol^{-1} and $L_L = 25$ kJ mol^{-1}, (e) $L_S = 25$ kJ mol^{-1} and $L_L = 25$ kJ mol^{-1}. Created using PANDAT [6].

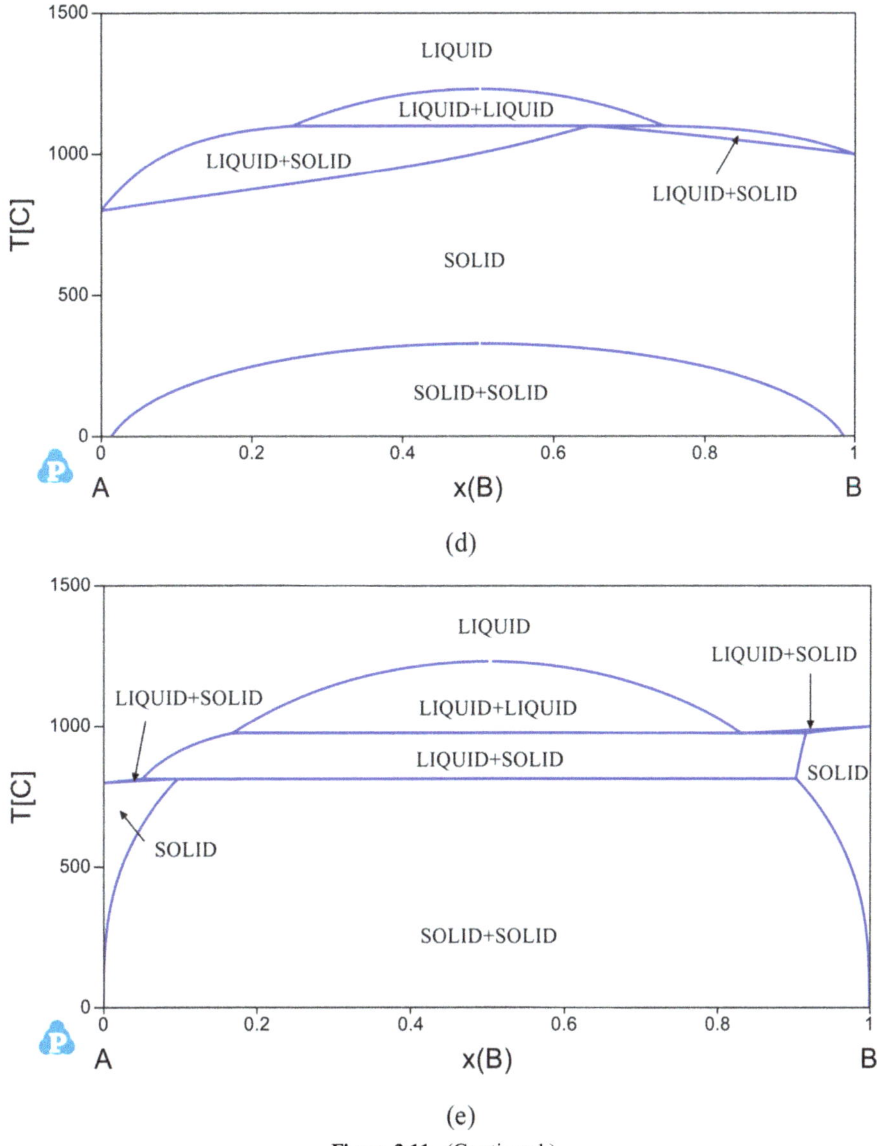

Figure 3.11. (Continued.)

composition ($x_B = 0.37$) and temperature ($T = 656\,°C$). A eutectic reaction is where a liquid transforms to two different solid phases at a given temperature, e.g. LIQUID → SOLID + SOLID. To explore the origins of the phase boundaries and the eutectic point in more detail, figure 3.12 plots the variation in Gibbs free energy of the two constituent phases at 656 °C and 750 °C. The eutectic temperature of 656 °C is chosen to illustrate why horizontal boundaries often appear in binary phase diagrams. It can be seen in figure 3.12(a) that at this temperature there is a common tangent that touches the energy curves at three points, rather than the usual two. At this temperature the

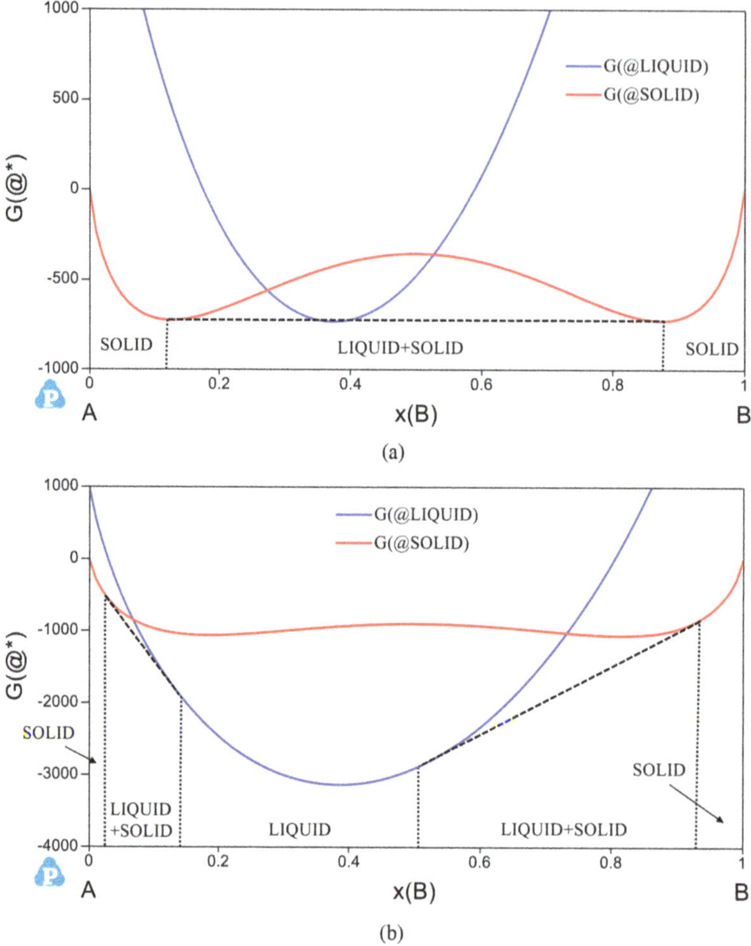

Figure 3.12. Gibbs free energy profiles for the eutectic binary of figure 3.11(c) at (a) the eutectic temperature of 656 °C, (b) 750 °C. Created using PANDAT [6].

LIQUID and SOLID phases just co-exist in the central region. The point at which the tangent touches the LIQUID curve is the eutectic composition. If the temperature is slightly lowered, the LIQUID curve moves upwards and loses contact with the tangent, leaving only the SOLID + SOLID phases. This sudden removal of the LIQUID phase from the region is the origin of the horizontal line. Figure 3.12(b) shows the energy profiles on the 750 °C isotherm, illustrated as a red dashed line on figure 3.11(c). There are two possible tangents that can be drawn between the LIQUID and SOLID energy curves, resulting in a pair of two phase regions. The other noticeable feature of this phase diagram which is very common is the appearance of the SOLID zones on the left- and right-hand sides of the diagram. Figure 3.12(b) shows that this occurs due to the sharp downward dip in the energy profiles on each side. This is because, as shown in figure 3.5, entropy always dominates the excess (mixing) energy at low compositions, meaning this dip always exists on the left- and right-hand sides. As the mixing energy increases to dissuade mixing, these regions can become smaller but they cannot be removed entirely. Figure 3.13 shows how the mole fraction of LIQUID and SOLID vary as the temperature is changed for two different nominal compositions. The alloy in figure 3.13(a) is known as a **hypo-eutectic** alloy, as its nominal composition ($x_0 = 0.2$) is below the eutectic composition ($x_0 = 0.37$). Interpreting the diagram from right to left, it can be seen that, as the temperature is decreased, the liquid phase smoothly reduces in fraction until it reaches the eutectic temperature (656 °C) at which point the remaining liquid instantly transforms to a solid (assuming equilibrium, i.e. the kinetics of the reaction are rapid). In figure 3.13(b) the eutectic composition ($x_0 = 0.37$) is chosen and shows that there is a sharp transition from LIQUID to SOLID + SOLID at the eutectic temperature, similar to that seen in pure metals.

Figure 3.11(d) now increases the miscibility gap in the liquid phase, forcing a LIQUID + LIQUID phase separation. Increasing the adversity to mixing in the solid phase as well, figure 3.11(e) shows the result when the SOLID + SOLID phase

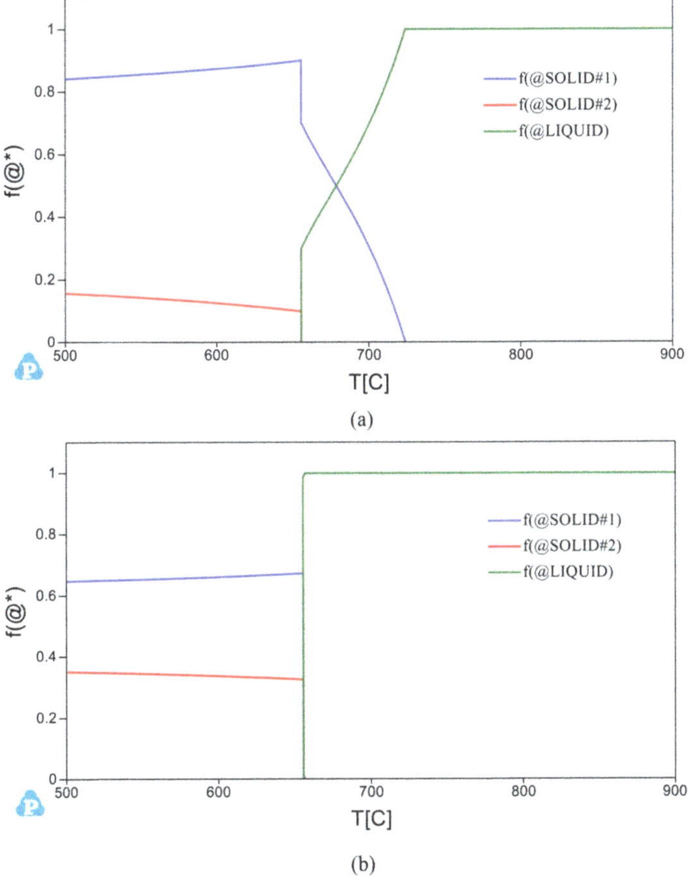

Figure 3.13. The mole fraction of the different phases for the eutectic binary phase diagram of figure 3.11(c) for a nominal composition of (a) $x_0 = 0.2$ and (b) the eutectic composition of $x_0 = 0.37$. Created using PANDAT [6].

separation region expands to impinge on the LIQUID + LIQUID region. This possesses many similar features to the Pb–Zn phase diagram, which can be downloaded from the NIMS database [4].

3.2.5 Solute partitioning and the partition coefficient

The previous sections have shown that: (a) most alloys solidify over a temperature range, and (b) for a given temperature in that range, the liquid and solid compositions are defined by the ends of the tie line isotherm. A consequence of this is **solute partitioning** during the solidification of non-eutectic alloys, leading to non-uniform distribution of the elements within the resulting solid [5]. This inhomogeneity can have a detrimental effect, as some areas of the microstructure will be weaker than others. To investigate this further, figure 3.14(a) shows an

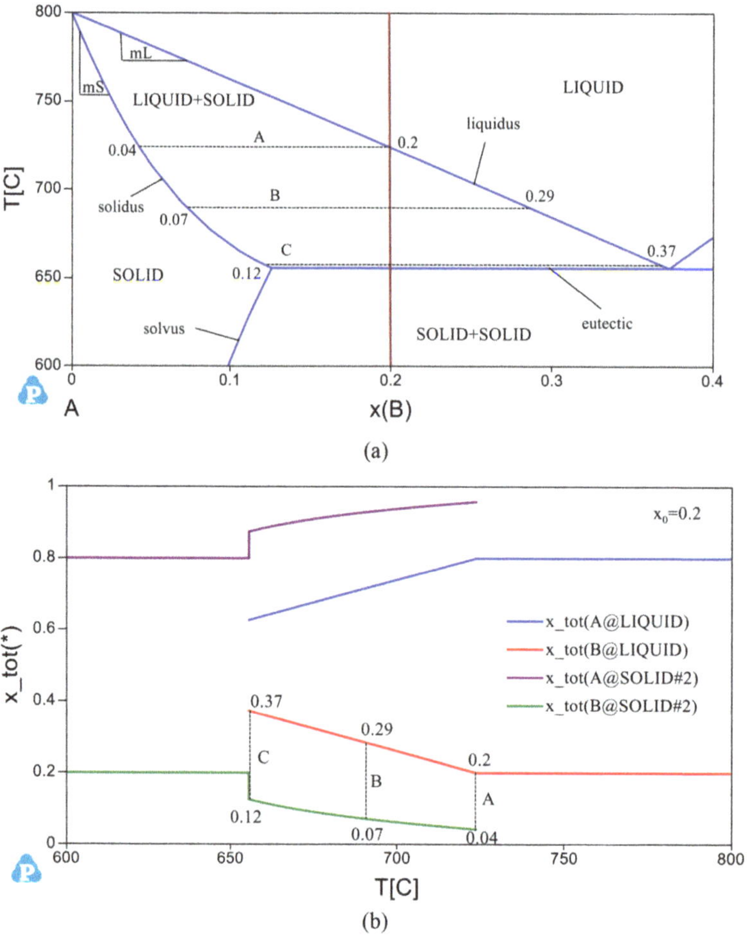

Figure 3.14. Solute partitioning during the equilibrium LIQUID to SOLID phase transition, (a) the composition in the two phases can be determined from the liquidus and solidus for the tie lines (A–C) shown, (b) these can then be plotted as a function of temperature. Created using PANDAT [6].

enlarged view of the left-hand LIQUID + SOLID region for the phase diagram in figure 3.11(c). Consider an alloy with nominal composition $x_0 = 0.2$, indicated by the vertical red line. Three tie lines are shown in this two phase region: (A) is at the onset of solidification, (B) is at the mid-point and (C) is near the final stages of the process. The **liquidus** and **solidus** phase boundaries lie at the end of these tie lines and give the compositions of B in the LIQUID and SOLID phases respectively, as labelled. It can be seem that the composition of the LIQUID phase is $x_B^L = 0.2$ at A and has increased to $x_B^L = 0.37$ at C. Likewise, the composition of B in the SOLID phase also increases from $x_B^S = 0.04$ at A to $x_B^S = 0.12$ at C. Therefore the solid that forms first has lower concentrations of B than the solid that forms last. As the temperature drops below the eutectic tie line, the last remaining 30% of LIQUID rapidly freezes at the eutectic composition. This **microsegregation** (over length scales of 10–100 μm) can be removed by annealing the microstructure at a high **solvus** temperature. Large scale **macrosegregation** also occurs in cast metals due to the temperature gradients that occur within the mould. This is less readily removed as the length scales are of the order of the cast size. Partioning can also be positively used to remove impurities from metals in a process known as **zone refining**.

The equilibrium segregation profile can be plotted as a function of temperature, as in figure 3.14(b), using PANDAT as follows. The tie lines from figure 3.14(a) have been superimposed to connect the two calculations.

PANDAT: plotting a segregation profile using 'twophase.tdb'
1. Ensure the TDB file 'twophase.tdb' is loaded with $L_s = 20$ kJ mol^{-1} and $L_L = 0$ kJ mol^{-1}.
2. Select 'PanPhaseDiagram → Solidification Simulation'.
3. Change 'x(A)' to 0.8 and 'x(B)' to 0.2 to define the nominal composition.
4. Change the 'Solidification model' to 'Equilibrium (Lever)'.
5. Uncheck the 'Start simulation from liquidus surface' and 'End when no more liquid' checkboxes.
6. Set the start temperature 'T(C)' to be '800' in the upper box.
7. Set the end temperature 'T_End [C]' to '600'.
8. Click on 'Options' and click on 'Default table'.
9. In the 'Choose table Columns' check the 'x_tot(*@*)' checkbox.
10. Uncheck any checked items in the 'Choose Y Axis Properties' column, and then check the 'x_tot(*@*)' checkbox.
11. Click 'OK' and the 'OK' again.

The propensity of an alloy for solute partitioning is defined by the **equilibrium partition coefficient**, which is the ratio of the equilibrium compositions in the solid and liquid phases

$$k_E = \frac{x_B^S}{x_B^L} \tag{3.20}$$

where $k_E > 0$. This is explored in more detail in the context of the LIQUID + SOLID two phase region shown in figure 3.14(a). Here we are assuming B is the **solute** (minority species) and A is the **solvent** (majority species). Although not strictly true, it is typically assumed that the liquidus and solidus are straight lines, defined on the left-hand side of the phase diagram by

$$T_L = T_A^M + m_L x_B^L \qquad\qquad T_S = T_A^M + m_S x_B^S \qquad (3.21)$$

respectively. The slopes of the liquidus and solidus, m_L and m_S, are shown in figure 3.14(a), and are defined by the thermodynamics. It is clear that at a given temperature ($T_L = T_S$) that

$$k_E = \frac{m_L}{m_S}. \qquad (3.22)$$

Derivation

Near the melting temperature of A, we can make the simplifying assumptions of $T \approx T_A^M$, $x_B^L \ll 1$ and $x_B^S \ll 1$. The two phases must satisfy the two equilibrium conditions defined by equation (3.17). To first order, the chemical potentials are defined from equation (3.10) as

$$\Delta\mu^S = \left.\frac{\partial g_S}{\partial x_B}\right|_{x_B = x_B^S} \approx L_S + RT_A^M \ln\left(x_B^S\right)$$

and from equation (3.11) as

$$\Delta\mu^L = \left.\frac{\partial g_S}{\partial x_B}\right|_{x_B = x_B^L} \approx g_L^B - g_L^A + L_L + RT_A^M \ln\left(x_B^L\right).$$

Therefore, one condition for equilibrium from equation (3.17) is that

$$\Delta\mu^S - \Delta\mu^L = g_L^B - g_L^A + L_L - L_S + RT_A^M \ln\left(\frac{x_B^S}{x_B^L}\right) = 0.$$

The equilibrium partition coefficient is therefore

$$k_E = \exp\left(\frac{g_L^B - g_L^A + L_L - L_S}{RT_A^M}\right) = \exp\left(\frac{\Delta S_B(T_B^M - T_A^M) + L_L - L_S}{RT_A^M}\right) \qquad (3.23)$$

using equation (3.12) and equation (3.20).

The second equilibrium condition in equation (3.17) gives the slopes of the liquidus and solidus. Expressing

$$\Delta\mu^S = \frac{g_L(x_B^L) - g_S(x_B^S)}{x_B^L - x_B^S}$$

to first order using equations (3.10) and (3.11) gives

$$L_S + RT_A^M \ln \left(x_B^S\right)$$
$$\approx \frac{[g_L^A(1 - x_B^L) + g_L^B x_B^L + L_L x_B^L - L_S x_B^S + RT_A^M(x_B^L \ln\left(x_B^L\right) - x_B^S \ln\left(x_B^S\right) + x_B^S - x_B^L)]}{(x_B^L - x_B^S)}.$$

Substituting $x_B^S = k_E x_B^L$ from equation (3.20) and multiplying out the denominator gives

$$x_B^L[RT_A^M(1 - k_E) + RT_A^M \ln\left(k_E\right) + g_L^A - g_L^B + L_S - L_L] = g_L^A.$$

Now equation (3.23) shows that $RT_A^M \ln\left(k_E\right) + g_L^A - g_L^B + L_S - L_L = 0$ so

$$x_B^L \approx \frac{g_L^A}{RT_A^M(1 - k_E)} = \frac{\Delta S_A(T_A^M - T)}{RT_A^M(1 - k_E)}.$$

Comparing this with equation (3.21) and equation (3.22) finally gives

$$m_L = -\frac{RT_A^M(1 - k_E)}{\Delta S_A} \qquad m_S = \frac{m_L}{k_E}. \tag{3.24}$$

The parameters for figure 3.14(a) are $\Delta S_A = \Delta S_B = 20$ J mol^{-1} K^{-1} and $T_A^M = 800\,°C$ and $T_B^M = 1000\,°C$. The values for these terms in the five cases shown in figure 3.11 are calculated in table 3.1. Cases (a), (b) and (e) are the same as $L_S = L_L$. The partition coefficient is greater than one, indicating that the concentration of the solute B is higher in the solid than the liquid, i.e. solute segregates to the solid phase. Case (d) has the largest partition coefficient, indicating that the driving force to segregate B (i.e. the value of $L_L - L_S$) is very large. Case (c) is the only case where the liquidus and solidus slope downwards, yielding a partition coefficient less than one. Here the solute B segregates to the liquid phase.

3.3 Sub-lattice models and stoichiometric phases

3.3.1 The sub-lattice model

The phase energies of crystalline solids are expressed using a sub-lattice model. Up to now we have only considered a single lattice, so it has not been apparent that we have been using a sub-lattice model. The sub-lattice model is only really apparent

Table 3.1. Liquidus and solidus slopes and equilibrium partition coefficient for the phase diagrams of figure 3.11.

Case	m_L (K)	m_S (K)	k_E
(a)	252	161	1.57
(b)	252	161	1.57
(c)	−372	−2235	0.17
(d)	3306	393	8.41
(e)	252	161	1.57

when elements in a crystal lattice are restricted to which atomic sites they can occupy. Two examples are shown in figure 3.15.

Figure 3.15(a) shows the austenitic (face-centred cubic, FCC) phase in the iron-carbon system. The Fe atoms occupy the substitutional FCC lattice sites, whereas the much smaller C atoms sit in the interstitial lattice sites. In this case, the phase is said to have two sub-lattices rather than one, the substitutional lattice and the interstitial lattice. The interstitial sites are mainly empty and hence this is not typically thought of as a lattice in the crystallographic sense. The thermodynamic model requires all sites to be occupied, so the empty sites are described as being occupied by vacancies. A new atomic element called VA for vacancies is introduced. This has zero mass with vacuum as the reference state and is treated just like any other element in the model. This is defined at the start of a TDB file by the following line.

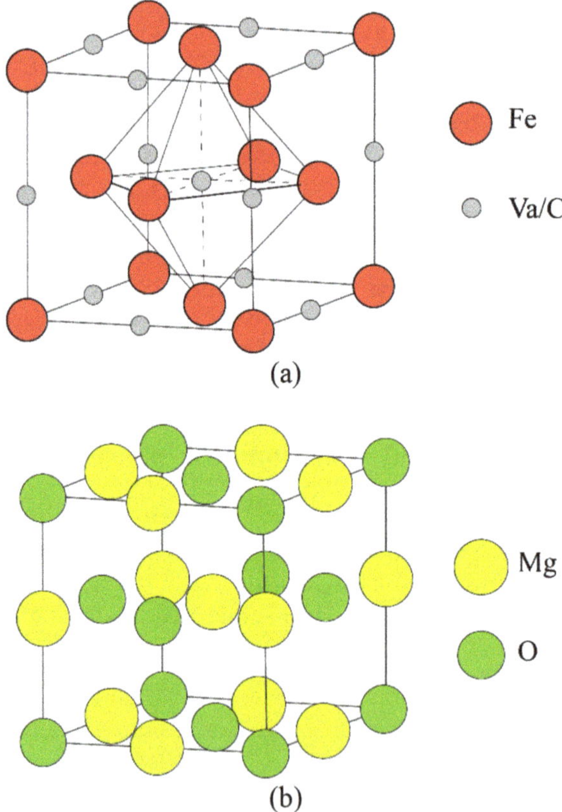

Figure 3.15. Crystal structures with two sub-lattices: (a) iron and carbon occupy separate sub-lattices. The majority of the interstitial carbon sites are vacant, i.e. occupied by vacancies. (b) magnesium and oxygen form a stoichiometric phase MgO where the Mg and O sit on similar but separate interpenetrating sub-lattices.

```
ELEMENT VA VACUUM 0.0 0.0 0.0 !
```

The FCC phase is then defined by the following lines.

```
PHASE   FCC   %   2  1  1  !
    CONSTITUENT FCC : FE : C,VA :   !
```

The first line indicates that the FCC phase consists of two sub-lattices (2), and that there is one site on the first sub-lattice (1) and one site on the second sub-lattice (1). The actual number of sites does not matter as long as the ratio is correct. The second line defines the elements that can occupy the FCC sub-lattices. The first sub-lattice (substitutional) can only be occupied by Fe, whereas the second sub-lattice (interstitial) can be occupied by C or VA.

If only one element can occupy a sub-lattice then the entropic contribution from that sub-lattice is zero, i.e. there is no other component for it to mix with. Hence there is no entropy arising from the Fe contribution, although in reality a small number of the Fe sites can also be vacant. There is an entropic term for the C and VA as these two 'components' can mix on the interstitial sub-lattice. The total energy per mole is now written in terms of the mole fractions of each component on a particular sub-lattice, denoted by y. The contribution arising from entropy is

$$g_{\text{ENTROPY}} = RT\big[y_C \ln(y_C) + y_{VA} \ln(y_{VA})\big]$$

where y_C and y_{VA} are the site fractions of carbon and vacancies on the interstitial sub-lattice respectively. This assumes there is 1 mole of substitutional (Fe) sites and 1 mole of interstitial sites. For the Fe sub-lattice all the sites are occupied by Fe so $y_{FE} = 1$. For the second sub-lattice we have $y_C + y_{VA} = 1$. The site fraction of carbon is related to the total mole fraction of carbon through $y_C = \dfrac{x_C}{x_{Fe}} = \dfrac{x_C}{1 - x_C}$, where the total number of interstitial sub-lattice sites is equal to the number of Fe sub-lattice sites for this stoichiometry, and hence the constraint that $0 \leqslant x_C \leqslant 0.5$ also follows.

Figure 3.15(b) shows the example of magnesium oxide, formed from the ions Mg^{2+} and O^{2-}. This has the strict **stoichiometric** formula of MgO, i.e. one O for every Mg. In this case, there are two sub-lattices, but only one component on each, Mg on one and O on the other. There is therefore no entropic contribution from either sub-lattice. The TDB file definition in this case uses the following line.

```
PHASE MgO   %   2 1 1 !
    CONSTITUENT MgO : Mg : O :   !
```

There are many different stoichiometric phases such as cementite (Fe_3C) or alumina (Al_2O_3). The TDB definition for cementite is as follows.

```
PHASE CEMENTITE   %   2 3 1 !
    CONSTITUENT CEMENTITE : FE : C :   !
```

The first line again shows there are two sub-lattice sites, the first sub-lattice has three sites and the second sub-lattice has one site. The second line indicates that only Fe can occupy the 3 sites of the first sub-lattice and only C can occupy the 1 site of the second sub-lattice. Again, there will be no entropic contribution from either sub-lattice in this case, as no mixing on the sub-lattices is possible.

3.3.2 Phase diagrams with stoichiometric phases

To illustrate the effect of a stoichiometric phase on the energetics of the system and the phase diagram, we now create an additional new phase with the stoichiometric formula AB_2 for the fictional thermodynamic model used so far. The composition of this phase is fixed and hence its energy (relative to the SOLID phase) is simply given by a single value

$$g_{AB2} = -1000 \text{ kJ mol}^{-1}$$

which may in reality also vary with temperature. Note that this defines the energy for one mole of the AB_2 molecule, which is equivalent to three moles of atoms (one mole of A atoms and 2 moles of B atoms). Therefore the energy is -1000 J per mole of AB_2, or a third of this, -333 J per mole of (A or B) atoms. To compare the energy of phases directly, the energy must be calculated in energy per mole of atoms. Hence the Gibbs free energy per mole for the AB_2 phase appears in the energetic calculation as -333 J mol^{-1}.

A phase with the formula AB_2 is created by adding the following lines to the end of the TDB file 'twophase.tdb' defined at the start of section 3.2:

```
PHASE AB2 % 2 1 2 !
  CONSTITUENT AB2 : A : B : !
  PARAMETER G(AB2,A:B) 298.15 -1000.0; 3000.0 N !
```

The effect of adding this phase to the eutectic phase diagram of figure 3.11(c) and the effect of the energy value of g_{AB2} are investigated in figure 3.16. It can be seen in

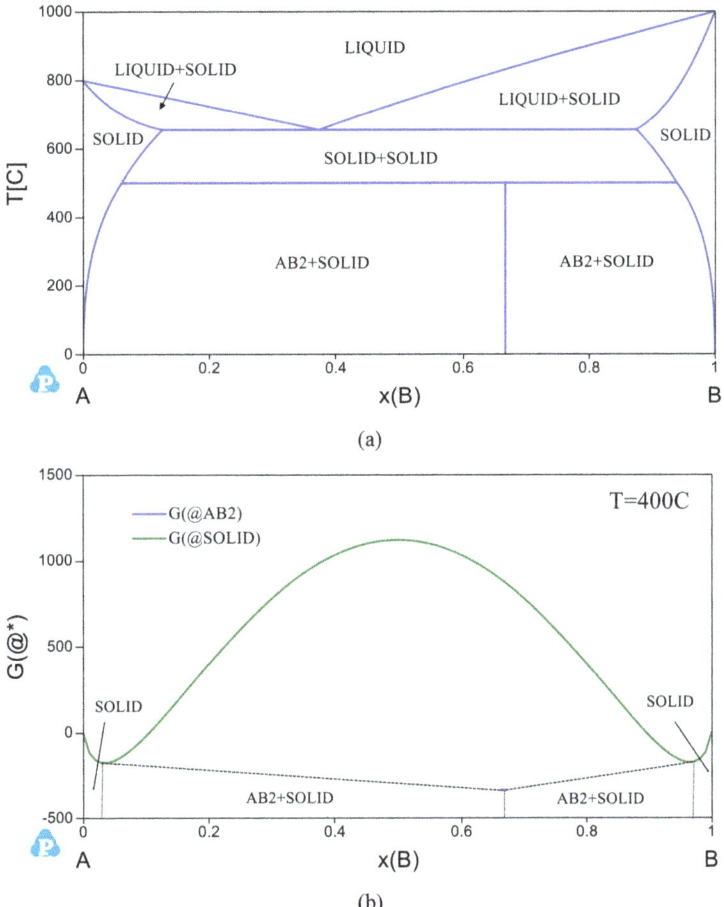

(a)

(b)

Figure 3.16. The eutectic phase diagram of figure 3.11(c) with an additional stoichiometric phase AB_2. (a) with $g_{AB2} = -1$ kJ mol^{-1}, (b) the associated phase energy profiles at 400 °C, (c) with $g_{AB2} = -5$ kJ mol^{-1}, (d) the associated phase energy profiles at 700 °C. Created using PANDAT [6].

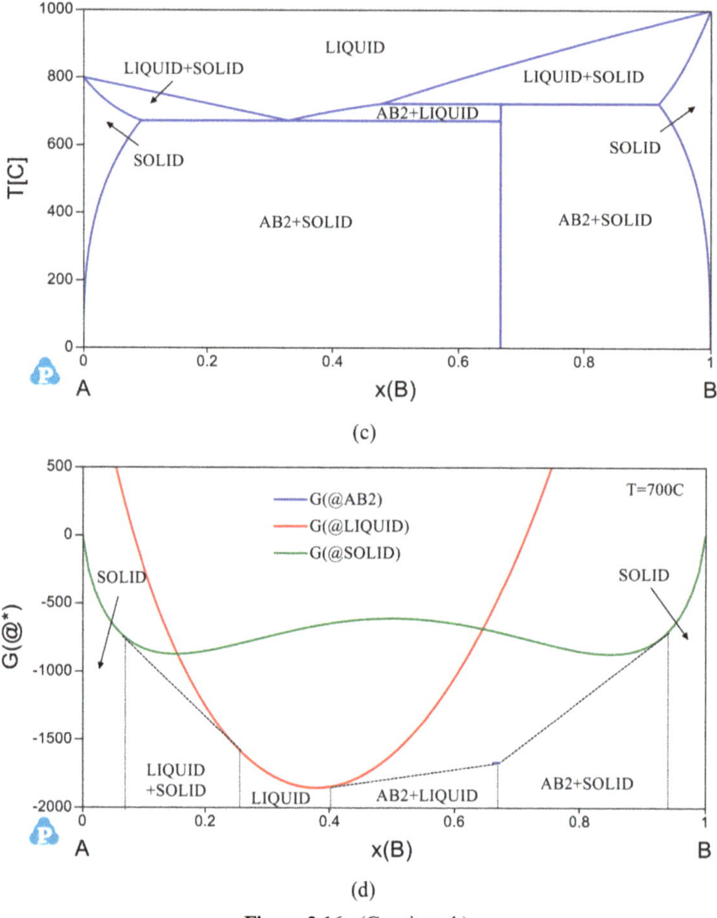

Figure 3.16. (Continued.)

figure 3.16(a) that the AB_2 phase appears at temperatures below 520 °C. The phase boundary between the two AB2 + SOLID regions is at the stoichiometric composition of $x_B^{AB2} = \dfrac{2}{1+2} = 0.67$ as expected. It is strictly vertical as the stoichiometric composition cannot change. The left-hand AB2 + SOLID region is a mix of AB_2 and the B-poor SOLID region on the left-hand side, and the right-hand AB2 + SOLID region is a mix of AB_2 and the B-rich SOLID region on the right-hand side. This can be seen in the cross-sectional Gibbs free energy profile at 400 °C, shown in figure 3.16(b). The energy of the AB_2 phase appears as a small (blue) line at the stoichiometric composition of $x_B^{AB2} = 0.67$ and has a value of $\dfrac{1}{3}g_{AB2} = -333$ J mol^{-1} as explained above. The energy of AB2 is now just a point, and hence the two phase tangent line can only be required to be tangential to the energy curves of the compositionally continuous phases. The minimum energy construction for determining the phase boundaries is shown. The two tangent lines from the SOLID can now achieve a lower energy state by forming a two phase region with the AB_2 point at this temperature.

Figure 3.16(c) illustrates the effect on the phase diagram of decreasing the energy of the AB2 phase to $g_{AB2} = -5000$ J mol^{-1}. This increases the favourability of forming this phase at higher temperatures, and pushes the vertical AB2 phase boundary further upwards, removing the SOLID + SOLID phase region and creating a AB2 + LIQUID region. The Gibbs free energy profile at 700 °C, shown in figure 3.16(d), provides another example of how the phase boundaries are determined for two phase regions which include a stoichiometric phase. As above, the energy of the AB2 phase appears at $x_{\mathrm{B}}^{\mathrm{AB2}} = 0.67$ and has a value of $\frac{1}{3}g_{AB2} = -1667$ J mol^{-1}.

3.4 Real binary systems

The main difference between the ficticious material systems considered so far and real systems is the complexity of the temperature-dependence of the energy coefficients and the number and variety of phases in the system. As discussed in section 1.4.1, TDB files for most binary material systems can be freely downloaded from the NIMS website [4]. The lead–tin (Pb–Sn) binary system is selected here for illustration purposes due to its relative simplicity. The phase diagram, shown in figure 3.17, is topologically very similar to the ficticious eutectic phase diagram in figure 3.11(c). The principal difference is that there are now three phases: one liquid phase (LIQUID) and two solid phases, one a face-centred cubic crystal (FCC_A1) and the other body centred tetragonal (BCT_A5). It is clear from the phase diagram that pure Pb is FCC_A1 and pure Sn is BCT_A5. The TDB file for this system is as follows.

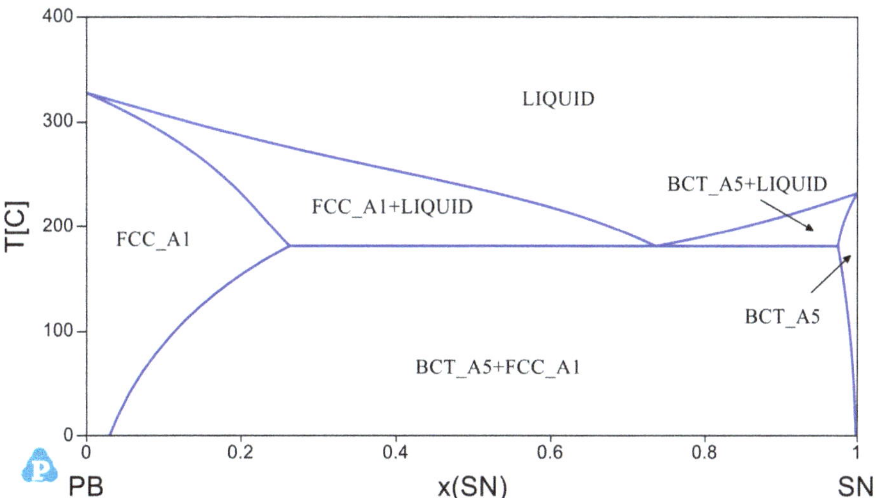

Figure 3.17. The phase diagram for Pb–Sn. Created using PANDAT [6].

Thermodynamic database file "PbSn.tdb" for the lead-tin binary system

```
$------------------------------------------------------------------
$ TDB file created by T.Abe, K.Hashimoto and Y.Sawada
$
$ Particle Simulation and Thermodynamics Group, National Institute
for
$ Materials Science. 1-2-1 Sengen, Tsukuba, Ibaraki 305-0047, Japan
$ e-mail: abe.taichi@nims.go.jp
$ Copyright (C) NIMS 2011
$
$ PARAMETERS ARE TAKEN FROM
$ A thermodynamic analysis of the Pb-Sn system and
$ the calculation of the Pb-Sn phase diagram,
$ T.L.Ngai, Y.A.Chang, CALPHAD, 5 (1981) 267-276.
$ ------------------------------------------------------------------
-----
 ELEMENT VA    VACUUM        0.0000E+00  0.0000E+00  0.0000E+00!
 ELEMENT PB    FCC_A1        2.0720E+02  6.8785E+03  6.4785E+01!
 ELEMENT SN    BCT_A5        1.1871E+02  6.3220E+03  5.1195E+01!
$
$ Standard Elemental Reference (SER) Enthalpies for Pb and Sn
$
 FUNCTION GHSERPB  298.15
   -7650.085+101.715188*T-24.5242231*T*LN(T)-3.65895E-3*T**2
   -0.24395E-6*T**3; 600.65 Y
   -10531.115+154.258155*T-32.4913959*T*LN(T)+1.54613E-3*T**2
   +805.644E23*T**(-9); 1200.00 Y
   +4157.596+53.154045*T-18.9640637*T*LN(T)-2.8829013E 3*T**2
   +0.098144E-6*T**3-2696755*T**(-1)+805.644E23*T**(-9); 5000.00 N !
  FUNCTION GHSERSN  100
   -7958.517+122.750027*T-25.858*T*LN(T)
   +5.1185E-04*T**2-3.192767E-06*T**3+18440*T**(-1); 250 Y
   -5855.135+65.427891*T-15.961*T*LN(T)-.0188702*T**2+3.121167E-
06*T**3
   -61960*T**(-1); 505.08 Y
   +2524.724+3.989845*T-8.2590486*T*LN(T)-.016814429*T**2
   +2.623131E-06*T**3-1081244*T**(-1)-1.2307E+25*T**(-9); 800 Y
   -8256.959+138.981456*T-28.4512*T*LN(T)-1.2307E+25*T**(-9); 3000 N
!
$
$ Function definitions
$

  FUNCTION GPBLIQ  298.15
    +4672.157-7.750257*T-6.0144E-19*T**7+GHSERPB#; 600.65 Y
    +4853.112-8.066587*T-8.05644E25*T**(-9)+GHSERPB#; 5000.00 N !
  FUNCTION GPBBCT  298.15   +489+3.52*T+GHSERPB#; 5000.00 N !
  FUNCTION GSNLIQ  100
    +7104.222-14.09088*T+1.49316649E-18*T**7+GHSERSN#; 505.06 Y
    +6970.705-13.813302*T+1.24912E+25*T**(-9)+GHSERSN#; 3000 N !
  FUNCTION GSNFCC  298.15   +4150-5.2*T+GHSERSN#; 3000 N !
 $
 $
```

```
$ Energy parameter definitions for LIQUID, FCC_A1 and BCT_A5 phases

 PHASE LIQUID:L % 1 1.0  !
   CONSTITUENT LIQUID:L :PB,SN :   !
   PARAMETER G(LIQUID,PB;0)         298.15 +GPBLIQ#; 5000 N !
   PARAMETER G(LIQUID,SN;0)         298.15 +GSNLIQ#; 4000 N !
   PARAMETER L(LIQUID,PB,SN;0)      298.15 +5125+1.46424*T; 6000 N !
   PARAMETER L(LIQUID,PB,SN;1)      298.15 +293.82; 6000 N !

 PHASE FCC_A1 % 2 1   1 !
   CONSTITUENT FCC_A1  :PB,SN : VA :  !
   PARAMETER G(FCC_A1,PB:VA;0)      298.15 +GHSERPB#; 5000 N !
   PARAMETER G(FCC_A1,SN:VA;0)      298.15 +GSNFCC#; 4000 N !
   PARAMETER L(FCC_A1,PB,SN:VA;0)   298.15 +5132.41+1.56312*T; 6000 N
 !

 PHASE BCT_A5 % 2 1   3 !
   CONSTITUENT BCT_A5  :PB,SN : VA :  !
   PARAMETER G(BCT_A5,PB:VA;0)      298.15 +GPBBCT#; 4000 N !
   PARAMETER G(BCT_A5,SN:VA;0)      298.15 +GHSERSN#;6000 N !
   PARAMETER L(BCT_A5,PB,SN:VA;0)   298.15 +17117.78-11.8066*T; 6000
 N !
```

The file has been slightly adjusted to enhance its readability. Some important features to note are that:

- lines starting with a '$' symbol are just comments.
- the opening comments acknowledge the authors, that the file has been created for the NIMS database [4] (reproduced with permission) and the source of the original journal reference for the thermodynamic parameters.
- the list of elements includes vacancies (VA) although, as we shall see, they are not actually used in this file.
- the Standard Elemental References (SER) for the contribution of the reference enthalpy to the Gibbs free energy of the elements, $H_{\mathrm{Pb}}^{\mathrm{SER}}$ and $H_{\mathrm{Sn}}^{\mathrm{SER}}$, are given in two rather long functions, GHSERPB and GHSERSN respectively. These are defined as functions of temperature over a number of different temperature ranges (note that T^{-9} is coded as 'T**(−9)', and ln (T) is coded as 'LN(T)' etc). As the name suggests, these reference functions are included as standard and are defined for all elements. For real materials, all other phase energies are defined in relation to these reference energies.
- a few function definitions follow. These generally just make the code easier to read and define some of the longer energy expressions for the pure Pb and Sn phases. Note that all these energies are added to the relevant SER energy for that element.
- finally the thermodynamic parameters for the three phases are defined, some using the above functions. In the LIQUID phase it is interesting that the

excess energy includes higher order interactions with two Redlich–Kister terms, L_L^0 and L_L^1. The FCC_A1 phase has two sub-lattices (as in figure 3.15(a)) where Pb and Sn can sit on the substitutional sub-lattice and the interstitial sub-lattice is entirely occupied by vacancies. The BCT_A5 phase also has two sub-lattices, where again Pb and Sn can sit on the substitutional sub-lattice and the interstitial sub-lattice is entirely occupied by vacancies. However, the definition shows that there are three times as many interstitial sites in the BCT_A5 phase. The code may have been set up to include these vacant sub-lattices to accommodate the inclusion of a third interstitial element when defining more advanced Pb–Sn alloys.

A final example of a phase diagram for a real material binary system is that for aluminium–platinum, shown in figure 3.18(a). Due to the number of regions, only the phase boundaries of the stoichiometric phases are labelled. As expected, six of these phases result in the formation of strictly vertical boundaries in the phase diagram due to their fixed composition. However, the boundaries for ALPT3 are not straight and there are two of them. To investigate why this is the case, it is instructive to look at the TDB code definition for this phase.

```
PHASE ALPT3% 3 .25 0.75 1 !
 CONSTITUENT ALPT3   :AL, PT : AL, PT : Va : !
```

The phase consists of 3 sub-lattices, with 0.25 sites on the first, 0.75 sites on the second and 1 site on the third. The second line shows that either Al or Pt can occupy the first or second sub-lattice. Therefore this phase is not strictly stoichiometric, although the phase name AlPt$_3$ suggests it is. The number ratio of the first two sub-lattice sites is 1:3 as expected for this formula. The nomenclature for this phase is slightly misleading, as it really implies that the first sub-lattice is just Al-rich, not 100% Al, and the second sub-lattice is Pt-rich, not 100% Pt. The third (interstitial) sub-lattice is fully occupied by vacancies and is largely irrelevant. The third sub-lattice therefore makes no contribution to the overall mole fraction, and the energy is defined for $0.25 + 0.75 = 1$ mole of (Al or Pt) atoms, i.e. a quarter of a mole of AlPt$_3$ molecules. The Gibbs free energy profiles of the different phases at 900 °C, shown in figure 3.18(b), confirm this conclusion. The six strictly stoichiometric phases are just a point in the energy landscape, whereas the energy for AlPt$_3$ is a continuous curve across the whole range of compositions. This shows why there are two ALPT3 boundaries and why they are not straight. Note that there is a kink in the AlPt$_3$ Gibbs free energy curve at the perceived stoichiometric composition of $x_{\mathrm{PT}}^{\mathrm{ALPT3}} = 0.75$.

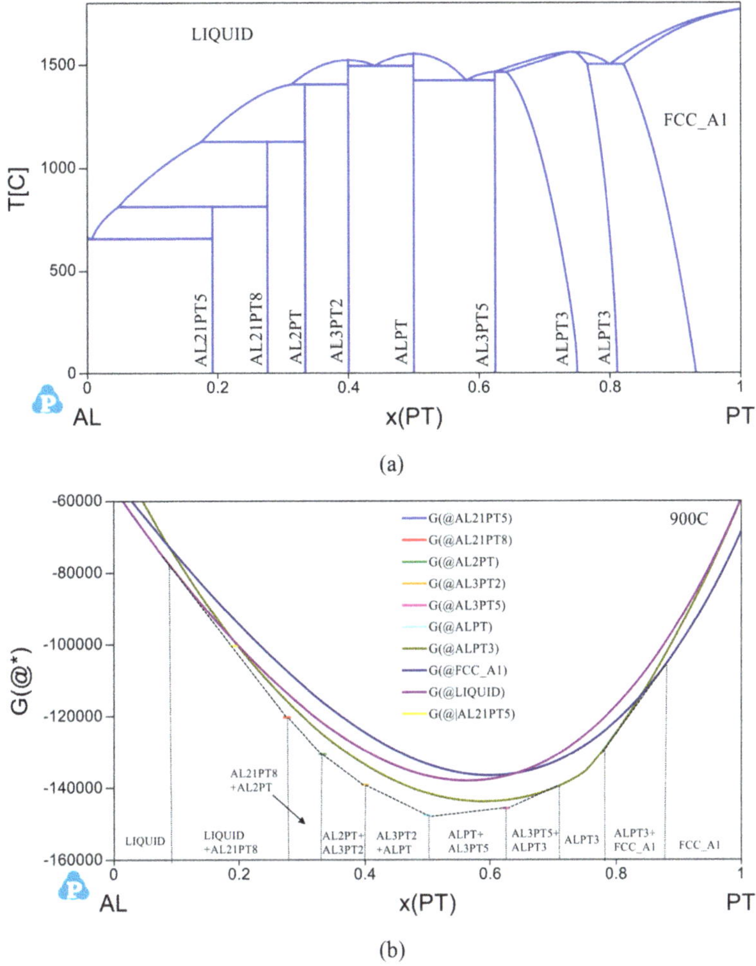

Figure 3.18. (a) The phase diagram for Al–Pt and (b) the Gibbs free energy profile along the 900 °C isotherm. The dashed line shows the minimum energy state in the two phase regions. The vertical dotted lines and labels indicate phase boundaries. Created using PANDAT [6].

References

[1] Porter D A, Easterling K E and Sherif M Y 2018 *Phase Transformations In Metals And Alloys* 3rd edn (Boca Raton, FL: CRC Press)

[2] Granta Designs Teach Yourself Phase Diagrams and Phase Transformations: www.grantadesign. com/download/pdf/edupack2015/Teach_Yourself_Phase_Diagrams_and_Phase_Transformations.pdf [accessed 1 October 2021]

[3] Campbell F C 2012 Phase Diagrams—Understanding the Basics (Materials Park, OH: ASM Int.)

[4] National Institute for Materials Science (NIMS) database: http://cpddb.nims.go.jp/cpddb/ periodic.htm [accessed 1 October 2021]

[5] Dissemination of IT for the Promotion of Materials Science (DoITPoMS) website: http:// doitpoms.ac.uk/tlplib/solidification_alloys/solute_partitioning.php [accessed 1 October 2021]

[6] Pandat: software suite for thermodynamic calculation and kinetic simulation of multi-component alloys. CompuTherm LLC, Madison, Wisconsin, USA. www.computherm.com [accessed 1 October 2021]

IOP Publishing

Thermodynamics, Kinetics and Microstructure Modelling

Simon P A Gill

Chapter 4

Ternary systems and beyond

It is possible to construct a phase diagram for three component (ternary) systems, but systems with more components cannot generally be interpreted graphically in this way. This chapter explains how to construct and interpret ternary phase diagrams, and then considers alloy systems with a higher number of elements.

4.1 How to read a ternary phase diagram

Ternary systems are introduced by extending the fictitious material system considered so far to three components: A, B and C. The compositions in ternary phase diagrams are projected onto the three sides of an equilateral triangle, as shown in figure 4.1(a). The corners represent each component, as indicated by the labels. Hence each side represents the concentration variation of a particular binary system, either A–B, A–C or B–C. The inside of the triangle represents all the ternary combinations. Each point inside the triangle, at (x_A, x_B, x_C), will represent a unique alloy composition. Lines of constant composition (of A, B or C) lie on lines that are parallel to the base of the triangle with the component of interest at the vertex. Three example lines for constant composition of A are shown in figure 4.1(a). For the $x_A = 0.6$ line, all the possible alloys on this line consist of 60 mol% A. The location of the 0.6 line is determined by the distance of the line from either the B or C vertices. At the vertex A we have 100% A, so the closer the line is to A the higher the composition of A. At B and C, the furthest points from A, the composition of A is 0%, as shown by the $x_A = 0$ line. Figure 4.1(b) shows three lines of constant composition, one for each component. Only two are necessary to define a particular alloy as $x_A + x_B + x_C = 1$, so knowing two values defines the other. All three lines of constant composition should cross at the same point if this relationship is satisfied correctly.

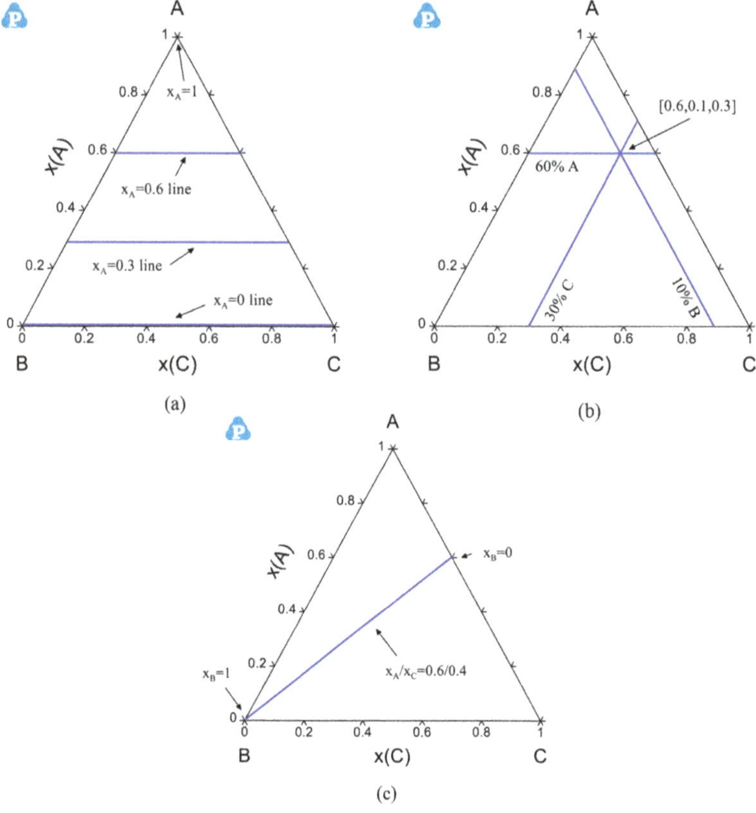

Figure 4.1. How to read a ternary phase diagram. (a) The contour lines for constant A are parallel to the base of the triangle where A is the vertex. The same applies for constant B and C lines. (b) Three contour lines of constant composition cross at a single point when defining the composition of a phase. (c) a line from the vertex (B in this case) has a constant ratio of the other components (60% A and 40% C). Created using PANDAT [3].

Another useful construction line on ternary phase diagrams is shown in figure 4.1(c). This goes from a vertex (B here) to the other side of the triangle. The composition of B goes from 100% (at vertex B) to 0% (at the other end), but at all times the remaining components (A and C) always have the same ratio. Here the A:C ratio is 60%:40%. The nearer the line is to the vertex A, the greater the proportion of A etc. Hence the composition on this line is $[0.6(1 - x_B), x_B, 0.4(1 - x_B)]$.

4.2 A simple ternary system

This example uses the following PANDAT example file ABC.tdb. Save the contents in a file of that name and load it into a new PANDAT workspace. Note that there are now three components (A, B and C) to be included.

A simple ternary thermodynamic database file: ABC.tdb

```
ELEMENT A    fcc                       0   0   0!
ELEMENT B    bcc                       0   0   0!
ELEMENT C    hcp                       0   0   0!

PHASE liquid  %  1 1 !
CONSTITUENT liquid  :A, B, C:  !
PARAMETER G(liquid,A;0)   298 0;    6000   N !
PARAMETER G(liquid,B;0)   298 0;    6000   N !
PARAMETER G(liquid,C;0)   298 0;    6000   N !

PHASE fcc  %  1 1 !
CONSTITUENT fcc  :A, B, C:  !
PARAMETER G(fcc,A;0)   298    -10000+10*T;  6000  N !
PARAMETER G(fcc,B;0)   298    -1000+10*T;   6000  N !
PARAMETER G(fcc,C;0)   298    -1100+10*T;   6000  N !

PHASE bcc  %  1 1 !
CONSTITUENT bcc  :A, B, C:  !
PARAMETER G(bcc,A;0)   298    -2000+10*T;   6000  N !
PARAMETER G(bcc,B;0)   298    -9000+10*T;   6000  N !
PARAMETER G(bcc,C;0)   298    -2500+10*T;   6000  N !

PHASE hcp  %  1 1 !
CONSTITUENT hcp  :A, B, C :  !
PARAMETER G(hcp,A;0)   298    -1500+10*T;   6000  N !
PARAMETER G(hcp,B;0)   298    -1600+10*T;   6000  N !
PARAMETER G(hcp,C;0)   298    -8000+10*T;   6000  N !
```

The file defines four phases: liquid (LIQUID), face-centred cubic (FCC), body-centred cubic (BCC) and hexagonal close packed (HCP). The equilibrium phase of pure A is FCC, pure B is BCC and pure C is HCP. The energy of pure A is $-10000 +10*T$ which melts when this is zero, at $T = 1000$ K. Similarly, the melting points of pure B and pure C are 900 K and 800 K respectively. Firstly we look at the three binary systems (A–B, A–C and B–C) to get some insight into the behaviour of the separate material systems, using a PanPhaseDiagram $\rightarrow$ Section Calculation. As shown in figure 4.2, these are all simple eutectic systems.

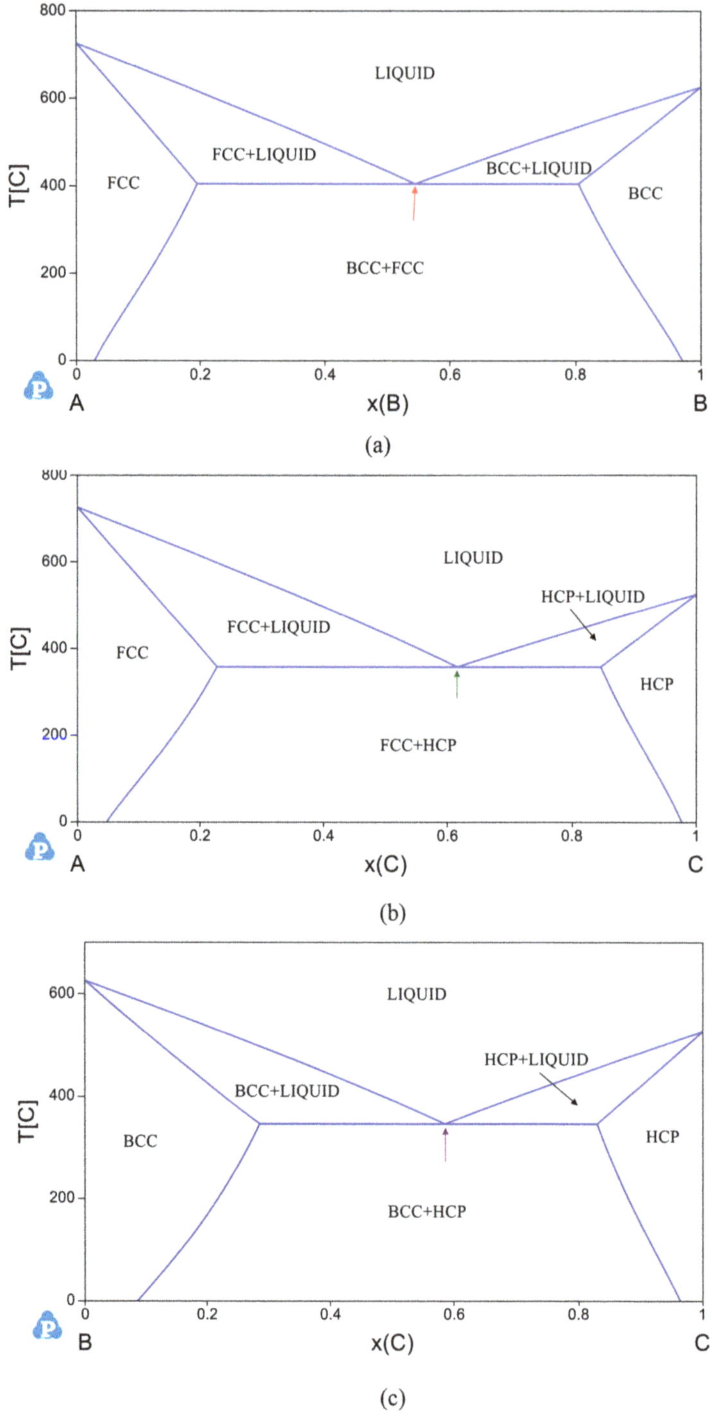

Figure 4.2. The three binary systems: (a) A–B, (b) A–C and (c) B–C. These three systems map onto the sides of the ternary triangle. The eutectic point in each system is indicated by a coloured arrow, reproduced on figure 4.3. Created using PANDAT [3].

It is not possible to plot the variation of all three components and temperature in a two-dimensional projection. Hence, in a ternary phase diagram, it is necessary to look at the phases at a single temperature, or look at the behaviour of the eutectic points (now lines) over the range of temperatures.

PANDAT: to plot the eutectic lines of a ternary phase diagram
1. Having ensured that 'ABC.tdb' is loaded, simply select PanPhaseDiagram → Phase Projection.
2. Check the 'Calculate Isotherms' checkbox, note that the 'Temperature Interval' is 50 °C and click 'OK'.
3. Hover the cursor over a contour line and press 'F2' to label it with its temperature value.

Figure 4.3(a) shows the projection of all the eutectic points in the system to form the blue lines shown on the ternary phase diagram. The triple junction of these lines near the centre indicates the lowest melting point in this system, just below 300° C. The coloured arrows indicate the positions of the respective eutectic points in the binary phase diagrams of figure 4.2. The isotherms indicate the position of the liquidus at the temperature shown. The values on the sides of the triangle should correspond with those observed in the binaries of figure 4.2. Comparing the liquidus surface of the ternary phase diagram to be a geographical feature, the temperature contours are similar to the height contours seen on a map, defining 'hills' that peak at the vertices A, B and C. The blue eutectic lines form at the bottom of the 'valleys' between the peaks, like 'streams'. The analogy fails at this point however, as all the 'streams' run down to the central triple point. The schematic diagram in figure 4.3(b) shows a 3D version of the ternary showing the 'peaks' and 'valleys', with the A–C binary visible on the visible side of the triangle. A nice video of a ternary being constructed can be found at [1]. PANDAT can also be used to create a 3D projection of the phase boundaries as follows.

PANDAT: to plot a 3D projection of a ternary phase diagram
1. In a PanPhaseDiagram workspace, go to Batch Calc → Batch Run.
2. Run the example file 'ABC_3D' in the PANDAT installation directory. The address is likely to be of the form '../Program Files/Computherm LLC/PANDAT 2020 DEMO/PANDAT 2020 Examples/PanPhaseDiagram/3D_Phase_Diagram/ABC_3D.pbfx' but will depend on the operating system you are using.

The output is shown in figure 4.3(c). It is difficult to visualise from a still image, but the advantage of plotting it in PANDAT is that you can rotate it using the mouse to see the construction from a variety of angles.

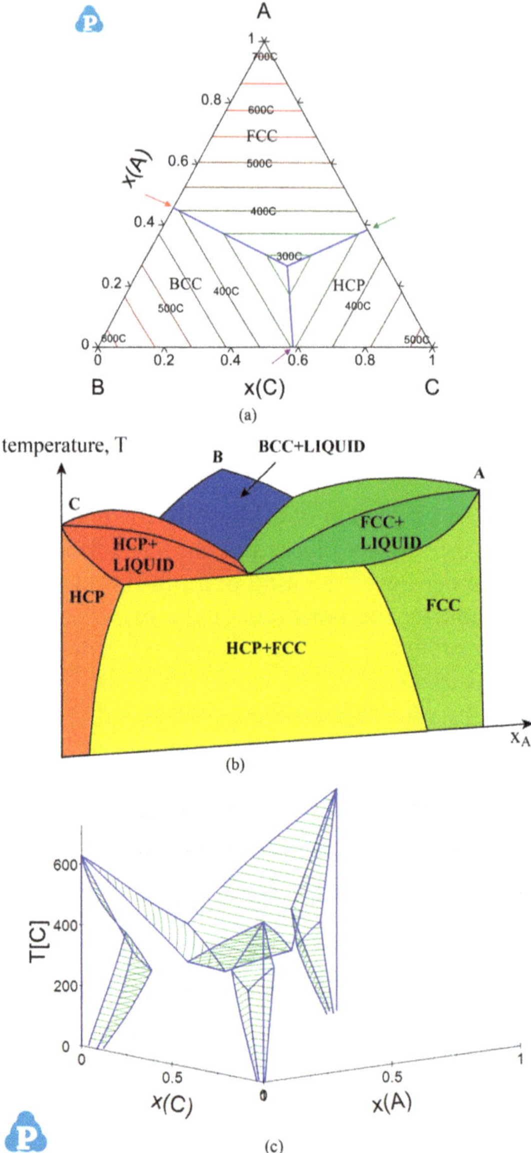

Figure 4.3. Phase projection for the A–B–C system. (a) This shows the eutectic lines between the different systems. The eutectic points of the three binary systems, indicated by arrows in figure 4.2, are also shown here. The composition of the liquidus at different temperature intervals is indicated by the contours. (b) A schematic drawing of the 3D projection of the temperature 'peaks' at the vertices A, B and C, and the eutectic 'valleys' that form between them. The lowest point is where the three 'valleys' meet. (c) The actual 3D projection, generated by the PANDAT example file 'ABC_3D.pbfx'. This can be rotated by a PANDAT user for better visualisation. (a) and (c) created using PANDAT [3].

Another useful plot is the ternary phases at a given temperature.

PANDAT: to plot ternary phases at a given temperature of 200 °C
1. Select PanPhaseDiagram → Section Calculation.
2. In the top left-hand box set 'T(C)' to 200, 'x(A)' to '1' and the other compositions to zero.
3. In the bottom left-hand box set 'T(C)' to 200, 'x(B)' to '1' and the other compositions to zero.
4. In the bottom right-hand box set 'T(C)' to 200, 'x(C)' to '1' and the other compositions to zero.
5. Click on 'Option' and 'Default table' and check the 'Show Tielines' checkbox.
6. Click 'OK' and then click 'OK' again.

Isothermal sections through the phase diagram at 200 °C and 300 °C are shown in figure 4.4. The green tie-lines highlight the two phase regions, on which levers rule (section 3.2.3) can be used in the normal way. There are no tie lines in the three phase region in the centre, as the concept does not function for more than two phases. At 200 °C the phases look very similar to those seen in the individual binaries of figure 4.2 until a large triangular three phase FCC+BCC+HCP region forms in the centre of the phase diagram. At 300 °C the centre of the phase diagram becomes more complicated as the liquid phase starts to appear. The liquid phase appears around the triple point in the centre, with two phase (solid+liquid) and three phase (solid+solid+liquid) regions radiating away from it.

Often it is of particular interest to determine the phase changes in a particular alloy as a function of temperature. As an example, the composition of (0.4,0.3,0.3) is selected. This alloy composition is indicated in figure 4.4 by a red star.

PANDAT: plotting phase changes as a function of temperature
1. Select PanPhaseDiagram → Line Calculation.
2. Set the left-hand parameters of 'T(C)' equal to '0', and the compositions of A, B and C to '0.4', '0.3' and '0.3' respectively.
3. Set the right-hand parameters of 'T(C)' equal to '500' and repeat the compositional settings above.
4. Click 'Option', and 'Default table' and ensure that 'f(@*)' is the only checkbox checked in the 'Choose Y Axis Properties' column.
5. Click 'OK' and 'OK' again.

The molar volume fraction of each phase as a function of temperature is plotted in figure 4.5. It is a simple way to assess the equilibrium composition of this alloy. There is a eutectic phase change at 274 °C, where there is a sudden change in the solid and liquid fractions. Below this, the solid is a mix of the FCC, BCC and HCP phases, with the FCC phase dominant. The HCP phase suddenly disappears at 274 °C, with the BCC and then

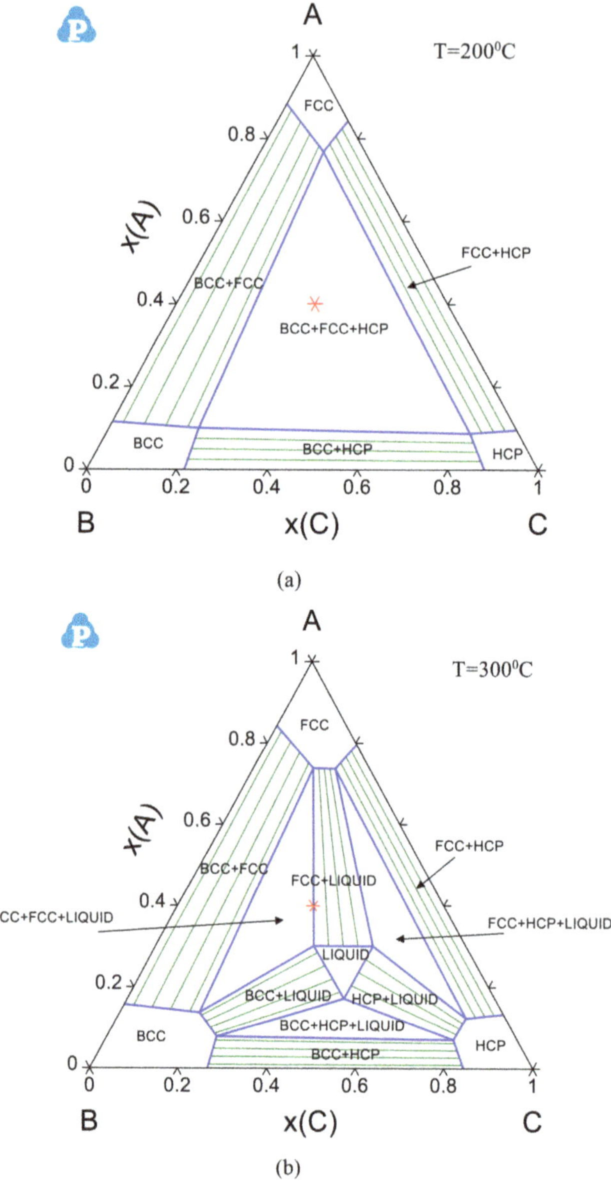

Figure 4.4. An isothermal cross-section of the A–B–C ternary at (a) 200 °C and (b) 300 °C. The red start indicates the composition (0.4,0.3,0.3) which is the subject of the line plot in figure 4.5. Created using PANDAT [3].

the FCC phases gradually phasing out as the temperature increases. The solid-to-liquid transition occurs between the temperature window of 274 °C to 369 °C.

In this simple ternary, there are no higher order thermodynamic interactions between the three components in any of the phases, i.e. the ternary is a simple linear combination of the three binary systems with no $x_A x_B x_C$ terms. This crude

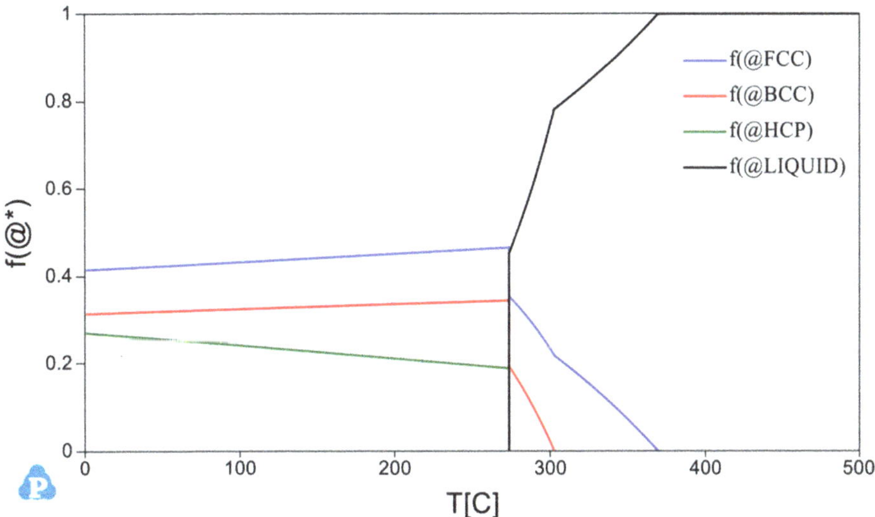

Figure 4.5. The molar fractions of each phase in the ternary system of composition (0.4,0.3,0.3) as a function of temperature. Created using PANDAT [3].

combination of three binaries does not generally describe the behaviour of the full ternary system very accurately. A few real ternary thermodynamic database files can be downloaded from the NIMS website for you to investigate. On the NIMS home page, shown in figure 1.5, select the 'Multi-component systems' link beneath the periodic table. Thirty different ternary TDB files are available including Ti–Al–V, Fe–Ni–Cr and Pb–Sn–Cu.

4.3 Examples with four or more components

Practical engineering alloys often use a complex mix of many elements to achieve the range of material properties desired. As introduced in section 1.4.1, commercial thermodynamic software packages such as Thermocalc, JMatPro and PANDAT can be used to analyse these systems. The free demo version of PANDAT that has been used so far is limited to three component systems. Hence the following two example calculations, performed in JMatPro, are included purely for reference. For multicomponent systems it is not possible to plot phase diagrams. Hence it is typical to use line plots to analyse the variation of phases and components as a function of temperature or composition, as in figure 4.5.

The first example is for the aluminium alloy 7075, widely used for structural components in aircraft due to its high strength-to-weight ratio. The elemental composition of the alloy additives (in weight percent) is given at the top of figure 4.6(a), with zinc being the primary alloying element and aluminium making up the (majority) remaining fraction. The plot here shows how the fraction of zinc (5.6 wt% of the alloy) is distributed between different phases in the alloy as temperature changes. At low temperatures nearly all the zinc is in the stoichiometric phase, MgZn2, where it would provide precipitate hardening, with it gradually

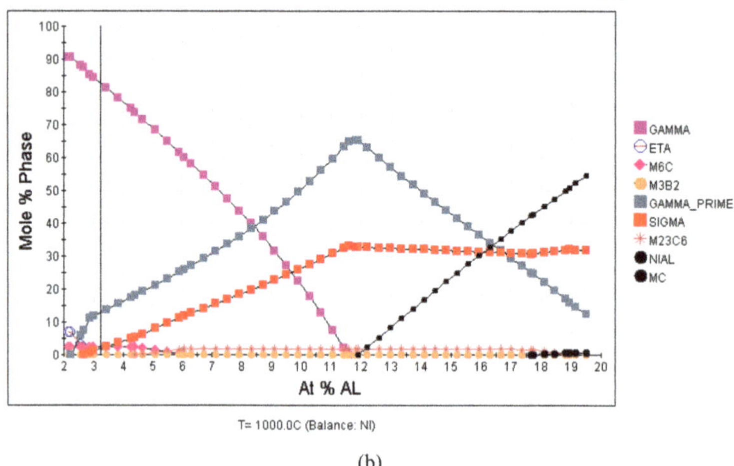

(a)

(b)

Figure 4.6. Two example thermodynamic calculations for commercial alloys created using the software JMatPro® [2]. (a) The zinc fraction in different phases in the aluminium alloy 7075 as a function of temperature. (b) the molar fraction of phases as a function of aluminium content for the nickel-based superalloy René 41 at 1000 °C.

going into solution in the Al matrix as the temperature increases to 390 °C, where the zinc provides solid-solution strengthening. The solid-to-liquid phase transition occurs over 105 °C starting at 525 °C. The right-hand legend indicates that there are 11 possible phases in this alloy.

The second example considers the nickel-based superalloy René 41. This alloy exhibits high-strength at temperatures in the 600 °C–1000 °C range, and is primarily used for high temperature, high stress applications such as gas turbine components. Again the composition is shown at the top of the diagram in figure 4.6(b). This alloy has a nominal aluminium content of 1 wt% (about 3.2 at%) but the calculation shows how the phase fractions change as this is varied from 2–20 at%. Al is a key element in the formation of the gamma prime strengthening phase, $Ni_3(Ti,Al)$, where Ti or Al can combine with Ni to form this phase. The gamma phase is the Ni matrix. The plot shows how the phase fractions change with Al content at 1000 °C. As the Al content is increased, it is no surprise to see the increase in the gamma prime phase at the expense of the gamma phase. The increase in gamma prime is beneficial to the strength of the alloy, but it can be seen that it is accompanied by the appearance of the sigma phase. This is known as a deleterious phase, which can be brittle and act as the initiation site for failure. Hence the nominal composition of René 41 (indicated on figure 4.6(b) by the vertical black line) has a relatively low Al content. The advantage of the relatively low gamma prime content of this nickel-based superalloy compared to others is that this increases the ductility of the material so that it can be formed into the shapes required. Note that, although predicted by the thermodynamics, the sigma phase often only appears after a very long time. This is because phase formation requires atoms to move from one region to another to happen, and this is controlled by kinetic processes such as diffusion. Therefore, to determine which phases are present at a given time requires the kinetics and thermodynamics to be considered together. This is the topic of the second part of this book.

References

[1] 3D animation showing the construction of a ternary phase diagram https://youtube.com/watch?v=doGhBC9sQlM [accessed 1 October 2021]

[2] JMatPro®, a software developed by Sente Software Ltd (UK) www.sentesoftware.co.uk/jmatpro [accessed 1 October 2021]

[3] Pandat: software suite for thermodynamic calculation and kinetic simulation of multi-component alloys. CompuTherm LLC, Madison, Wisconsin, USA www.computherm.com [accessed 1 October, 2021]

IOP Publishing

Thermodynamics, Kinetics and Microstructure Modelling

Simon P A Gill

Chapter 5

Driving force for nucleation and growth

The thermodynamics of phase change has so far assumed that a new phase will appear if it lowers the overall chemical Gibbs free energy of the system. How the new phase appears and how other energetic contributions affect the system have not been considered. In practice, a new phase will appear by the formation of small clusters of atoms. The initial atomic clusters are called **nuclei**, and the process of their appearance is known as **nucleation**. If nuclei are energetically favourable they will then **grow** by the attachment of additional atoms. A solid nucleus growing from a liquid will typically form a grain, and a solid nucleus growing within a solid will usually form a precipitate. There is an additional energy penalty for the formation of nuclei due to the **interfacial energy** between the nuclei and the host phase. This is not accounted for in the phase diagrams that have been considered so far. To initiate nucleation of a solid in a liquid, the interfacial energy means that a temperature drop below the liquidus, known as **undercooling**, is needed to supply additional phase energy to overcome this energy penalty. In addition, the interfacial energy depends on the size of the nuclei, and this introduces a microstructural length scale into the system for the first time. In this chapter, the phase energy change due to nucleation is introduced in sections 5.1 and 5.2, and then the interfacial energy is accounted for in section 5.3.

5.1 Phase energy change for nucleation and growth

Consider the sudden appearance of a small SOLID particle within the LIQUID phase. For the single component material system of chapter 2 the phase energy change (per mole of SOLID) is given by

$$\Delta g_{L \to S}(T) = g_S(T) - g_L(T) = -L_f \frac{\Delta T}{T_M} \tag{5.1}$$

where the last equation assumes the approximation of equation (2.6) and $\Delta T = T_M - T$ is the **undercooling**. This shows that when $\Delta g_{L \to S} > 0$ the energy increases so the LIQUID-to-SOLID phase change is opposed, i.e. $T > T_M$. When $\Delta g_{L \to S} = 0$ we are at the LIQUID–SOLID phase boundary, i.e. $T = T_M$. Only when the undercooling is positive, $\Delta T > 0$, is energy released to drive the phase change, i.e. $\Delta g_{L \to S} < 0$. Equation (5.1) shows that the energy release driving phase change increases with the undercooling, i.e. distance from the phase boundary.

We now extend this analysis to our fictitious binary system of component A and B of chapter 3. The calculation of the phase energy change may appear to be simply the difference between the equilibrium energies of the new SOLID phase and the LIQUID phase it has replaced, i.e. $g_S(x_B^S, T) - g_L(x_B^L, T)$. However, it is clear from two-phase energy plots, such as that in figure 3.8, that $g_S(x_B^S, T) \neq g_L(x_B^L, T)$ at a phase boundary, and in fact $g_S(x_B^S, T) > g_L(x_B^L, T)$ in the two phase region, where the SOLID is expected to appear. This approach does not work as the LIQUID composition cannot reach x_B^L until the equilibrium volume fraction of SOLID, f, has been reached. To calculate the energy change correctly we must account for the changes in composition of the LIQUID during phase change.

Derivation of chemical driving force for nucleation

Consider a binary alloy with components A and B, where the nominal composition of B is x_0. Before nucleation the initial energy of the system, measured in Joules, is

$$G_L = g_L(x_0)N_0$$

where N_0 is the number of atomic moles of liquid. A SOLID precipitate of N_S moles is now assumed to form in the LIQUID with equilibrium composition, x_B^S. This is illustrated in figure 5.1, where in this case $x_B^S < x_0$, so some of the solute (B) is ejected from the nucleus when it forms (and as it grows). Here the LIQUID parent phase is assumed to be homogeneous such that the original

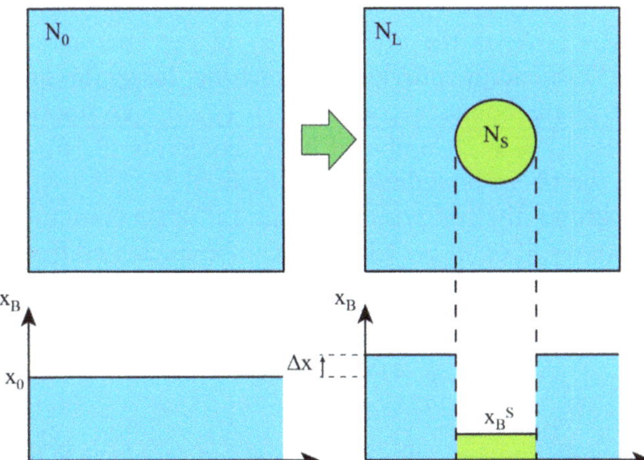

Figure 5.1. A volume of N_0 moles of LIQUID at composition x_0 (left) transforms to create a SOLID particle of N_S moles at concentration x_B^S (right). The compositional variation across the sample is shown below. As $x_B^S < x_0$ in this case, the excess B material is ejected from the nucleus into the surrounding liquid to increase the average LIQUID composition to $x_0 + \Delta x$.

concentration x_0 uniformly rises to a slightly higher concentration $x_0 + \Delta x$. (Note that in practice, equilibrium conditions will be maintained at the interface causing a local enrichment of B in the LIQUID local to the nucleus. This is considered further in section 7.2.2).

The phase energy of the two phase system after nucleation is therefore

$$G_{L+S} = g_S(x_B^S)N_S + g_L(x_0 + \Delta x)N_L$$

where N_L is the number of moles of liquid remaining after nucleation. For conservation of the total number of moles it is required that

$$N_S + N_L = N_0. \tag{5.2}$$

The mole fraction of component B (and therefore A also) must also be conserved, such that

$$x_B^S N_S + (x_0 + \Delta x)N_L = x_0 N_0. \tag{5.3}$$

Now, assuming Δx is small, we write

$$g_L(x_0 + \Delta x) \approx g_L(x_0) + \Delta\mu^L(x_0)\Delta x$$

where the definition of chemical potential difference in equation (3.8) has been used. The energy change is therefore

$$\Delta G_{L\to S} = G_{L+S} - G_L = g_S(x_B^S)N_S + g_L(x_0)N_L + \Delta\mu^L(x_0)\Delta x N_L - g_L(x_0)N_0.$$

Defining N_L using equation (5.2) and Δx using equation (5.3) gives

$$\Delta g_{L\to S}^{nuc} = g_S(x_B^S) - [g_L(x_0) + \Delta\mu^L(x_0)(x_B^S - x_0)] \tag{5.4}$$

where $\Delta g_{L\to S}^{nuc} = \Delta G_{L\to S}/N_S$ is the phase energy change per mole of transformed (SOLID) phase. This energy provides the **driving force for nucleation**. At the LIQUID–SOLID phase boundary, where $T = T_M$, the composition of B in the LIQUID phase is $x_B^L = x_0$. Combined with equation (3.16), this shows that $\Delta g_{L\to S}(T_M) = 0$ at the phase boundary as required.

As the nucleus grows, the composition of the LIQUID continues to change. The non-equilibrium composition of the LIQUID can be calculated from equation (3.14) such that

$$\tilde{x}_B^L = \frac{\left(x_0 - \tilde{f}_S x_B^S\right)}{1 - \tilde{f}_S} \tag{5.5}$$

where $\tilde{f}_S = N_S/N_0$ is the (non-equilibrium) mole fraction of the SOLID phase, and $\tilde{x}$ is the (non-equilibrium) composition in the LIQUID. The general **driving force for nucleation and growth** can be determined from equation (5.4) by substituting the updated composition of the LIQUID phase $\tilde{x}_B^L$ for the nominal composition x_0 to give

$$\Delta g_{L \to S} = g_S\left(x_B^S\right) - \left[g_L\left(\tilde{x}_B^L\right) + \Delta \mu^L\left(\tilde{x}_B^L\right)\left(x_B^S - \tilde{x}_B^L\right)\right]. \tag{5.6}$$

Nucleation occurs when $\tilde{f}_S = 0$ and so this reduces to equation (5.4) as required. As growth proceeds at the undercooled temperature, $T = T_M - \Delta T$, the liquid composition will continue to increase until the equilibrium composition has been reached. At this point $\tilde{x}_B^L = x_B^L$ and $\tilde{f}_S = f_S$ and equations (3.14) and (5.6) show that $\Delta g_{L \to S} = 0$. This demonstrates that the driving force for growth reduces as equilibrium is approached and is zero once equilibrium is reached, as expected. This is therefore a vital expression for modelling the appearance and growth of phases. To be fully utilised in a microstructural model, it must be combined with the kinetics of material transport, and this is considered in chapter 8.

5.2 Calculating the phase energy change

The phase energy change of equations (5.4) and (5.6) can be determined from a graphical construction using the phase energy profiles. This is illustrated for the nucleation energy of equation (5.4) using the thermodynamics of the eutectic phase diagram produced in figure 3.11(c). Consider an alloy with a nominal composition of $x_0 = 0.2$ at a temperature of $T = 700\,°C$. The relevant left-hand LIQUID + SOLID two phase region of interest, where nucleation of the SOLID is expected, is reproduced in figure 5.2(a).

The Gibbs free energy profiles of the two phases at 700 °C are given in figure 5.2(b), which illustrates the construction process to determine the energy of phase change. The process is as follows: (1) draw the equilibrium common tangent line to the two phases; (2) identify the equilibrium composition of B in the SOLID where the tangent touches the SOLID curve and draw the vertical line $x_B = x_B^S$. This touches the SOLID line at $g_S(x_B^S)$; (3) draw the vertical line $x_B = x_0$. This touches the LIQUID line at $g_L(x_0)$; (4) Draw a tangent to the LIQUID curve at $x_B = x_0$. This has slope $\Delta \mu^L(x_0)$ and intercepts the $x_B = x_B^S$ line at $g_L(x_0) + \Delta \mu^L(x_0)(x_B^S - x_0)$; (5) the vertical distance between this intercept and the SOLID curve is the phase change energy $\Delta g_{L \to S}$, indicated by the green arrow.

The value of the phase change energy can be found by placing the mouse at either end of the green arrow and recording the value of the energy at this point (shown in the bottom left corner of the PANDAT window). For the drawing in figure 5.1(b) these values are: $g_S(x_B^S) = -677\,J\,mol^{-1}$, $g_L(x_0) + \Delta \mu^L(x_0)(x_B^S - x_0) = -182\,J\,mol^{-1}$. The phase change energy is the difference between these two such that $\Delta g_{L \to S} = -495\,J\,mol^{-1}$.

Although the graphical construction provides useful insight into the origin of this energy difference, it is not a very accurate method. It can of course also be calculated more accurately numerically. In PANDAT this is known as the driving force for phase change, denoted by the variable name DF.

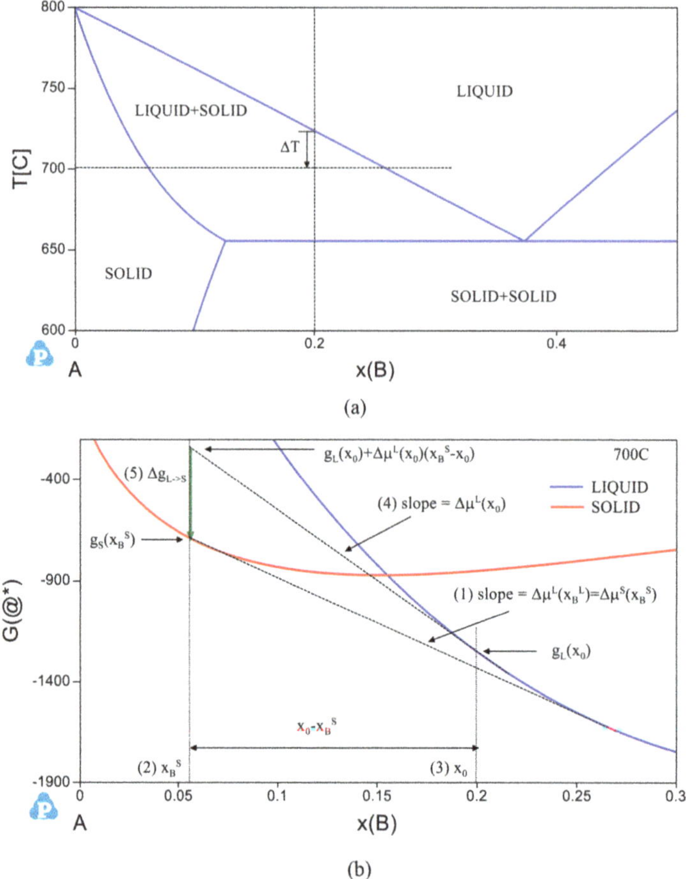

Figure 5.2. (a) The left-hand LIQUID + SOLID region of the eutectic phase diagram of figure 3.11(c) with $L_S = 20\ 000$ J mol^{-1} and $L_L = 0$. Nucleation in an alloy with $x_0 = 0.2$ at a temperature of 700 °C where the undercooling $\Delta T = 22$ °C. (b) the Gibbs free energy profiles at 700 °C showing the graphical construction of the phase transition energy $\Delta g_{L\to S}$. Created using PANDAT [1].

PANDAT: calculating the energetic driving force, DF

1. Start a new PanPhaseDiagram workspace and load the 'twophase.tdb' database file with $L_S = 20\ 000$ J mol^{-1} and $L_L = 0$.
2. Select PanPhaseDiagram → Point Calculation.
3. Set 'T(C)' to '700', 'x(A)' to '0.8' and 'x(B)' to '0.2' to define the point of interest on the phase diagram.
4. Click on 'Select Phases'. The SOLID and LIQUID phases should appear as 'Entered Phases'. These are the phases in the equilibrium calculation. To calculate the driving force, the SOLID phase must be added to 'Dormant Phases', i.e. it is not at equilibrium with the LIQUID phase. To do this, click on 'SOLID' in the 'Entered Phases' column and click on the blue right arrow button to the left of this column to move it into the 'Dormant Phases' column. Click 'OK'.

5. Now click on 'Options', check the checkbox next to the variable 'DF(@!*)' in the 'Choose table Columns' column. Click 'OK' and 'OK' again.
6. Note: do not confuse 'DF(@!*)' with the variable called 'DF(@|*)' which uses a | rather than a !. The latter finds the driving force for an 'Entered' phase, which necessarily means that this is only when the phase is not in equilibrium and therefore not expected to appear, i.e. when the driving force is negative as it is not energetically favourable.

The result will appear in the Graphics window. Scroll down and you will see that for the SOLID phase the value of 'DF (J mol^{-1})' is given as 496.72 J mol^{-1}. This is very close to the above value of -495 J mol^{-1}, although of a different sign. This is because energy release ($\Delta g_{L\rightarrow S}<0$) results in a positive driving force for phase change.

It is possible to plot contours of driving force on a phase diagram.

PANDAT: plotting the driving force, DF, on a phase diagram
1. For this example, create a new PanPhaseDiagram workspace and load the 'PbSn.tdb' file from section 3.4. (If need be, you can create the file by cut-and-pasting the contents from this document into a text file of that name.)
2. Plot the binary phase diagram as usual. The Graph and Table that produced this plot will appear in the Workspace window. If this is the first plot it will be called 'section'.
3. There are two solid phases in this system: FCC_A1 and BCT_A5. We will plot the driving force for the appearance of the BCT_A5 phase.
4. Go to PanPhaseDiagram → Section Calculation. Click on 'Select Phases' and move the BCT_A5 phase into the 'Dormant Phases' column using the blue arrows (as described above). Click 'OK'.
5. Click on 'Contour Lines' and then 'Add' to create a new set of contours. In the Contour Type box delete the text 'f(@*)' and type 'DF(@!BCT_A5)' instead. Change the 'Start' value to '-2000', the 'Stop' value to '2000' and the 'Step' value to '200'. Click 'OK' and 'OK' again to perform the calculation. The new phase diagram will be called 'section_1'.
6. The contours appears as red lines, but the phase boundaries are no longer correct as the BCT_A5 phase is dormant, i.e. not included in the equilibrium calculation.
7. Now plot the contours on the first copy of the phase diagram, called 'section'. Click on the original graphs tab to display the proper phase diagram on screen. Select to 'Graph → Edit Plots'. Click on 'Import a table File'. Move out of the 'section' directory and into the 'section_1' directory. Open the 'Table' sub-directory, select the 'Contour_DF(@!BCT_A5)' table and click 'Open'.
8. The data from the contour plot should appear in the 'Setup Plot' window in the left-hand 'Available Columns' column. Use your mouse to drag the 'x(Sn)[mole/mole]' data into the first empty box in the 'X Axis' column in the 'New Plot' window. Do the same for the 'T[C]' data into the 'Y Axis' column. Click 'OK' and the red contour lines should appear on the original phase diagram.

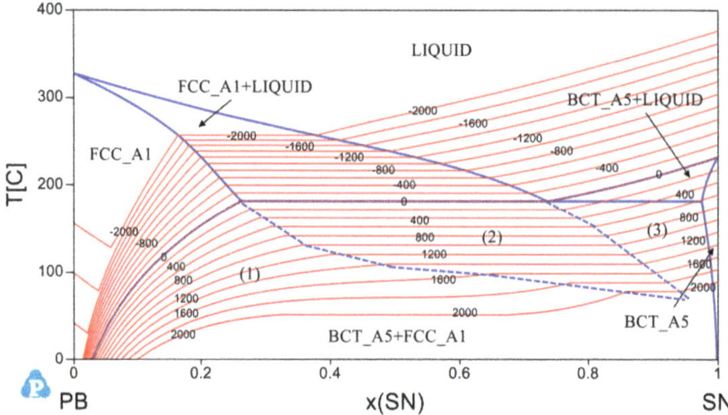

Figure 5.3. Contours of driving force for the nucleation of the BCT_A5 phase in the Pn–Sn binary phase diagram. Created using PANDAT [1].

The result is shown in figure 5.3. The line of zero driving force defines the phase boundaries between regions where the BCT_A5 does exist at equilibrium and where it does not. The driving force is negative where BCT_A5 does not exist at equilibrium and positive where it does exist. The driving force increases with distance from the phase boundary. The blue dotted lines have been added to highlight where kinks in the contour lines occur, delineating extended non-equilibrium phase boundaries in the case that the BCT_A5 phase did not nucleate, e.g. due to rapid cooling. The FCC_A1 region extends into (1), the FCC_A1 +LIQUID phase extends into (2) and the LIQUID phase extends into region (3).

5.3 The total driving force for nucleation

The previous sections in this chapter have investigated the phase energy change that occurs upon nucleation of a new phase at an undercooling ΔT. In energetic terms, the undercooling is required to overcome the additional energy required to create the interface between the host phase and the nucleus. This additional energy is described by

$$G_{\text{int}} = \gamma_{\text{int}} A_{\text{int}} \tag{5.7}$$

where A_{int} is the area of the interface and γ_{int} is the energy per unit area of the interface (J m^{-2}). The latter is a property of the two phases which it forms an interface between. It is typically assumed to be independent of temperature but can be a function of the orientation of the interface. This effect is not considered here and an average interfacial energy is assumed. This is not always easy to define with great accuracy, with values typically in the range 0.01 J m^{-2} to 1 J m^{-2} being common. The origin of the interfacial energy is similar to that for the energy for other defects, such as those considered in section 1.2.3, e.g. grain boundaries. The atomic packing at the interface is different to that which minimises the energy within the two constituent phases, and therefore is a region of higher energy. The interface is typically 1–2 atomic distances thick.

The introduction of the interfacial area means that it is necessary to convert molar phase energies to volumetric phase energies. As such, the molar volumes of the two phases, Ω_L and Ω_S, are introduced respectively. These are the volumes occupied by 1 mole of the phase. These are constants, and changes due to thermal expansion and composition are commonly neglected. They are typically of the order of 10^{-5} m^3 mol^{-1}. The phase energies of our example LIQUID and SOLID phases per unit volume are therefore g_L/Ω_L and g_S/Ω_S, measured in J m^{-3}. In general it is reasonable to assume to first order that $\Omega = \Omega_S = \Omega_L$.

Using equation (5.4), the total energy change (Joules) upon formation of a nucleus is therefore

$$\Delta G_{L\rightarrow S}^{nuc} = \frac{\Delta g_{L\rightarrow S}^{nuc}}{\Omega} \cdot V_{nuc} + \gamma_{int} A_{int} \tag{5.8}$$

where V_{nuc} is the volume of the nucleus. Note that, for a positive undercooling, the phase energy change is always negative and the interfacial energy is always positive. There is therefore competition between the two terms, with the phase energy promoting nucleation as the undercooling increases, and the interfacial term acting to suppress it.

5.3.1 Homogeneous nucleation

Homogeneous nucleation is the case where a nucleus has an equally likely chance of appearing at any point within the host phase (taken to be LIQUID here), i.e. there are no locations that are more energetically favourable than others. In this case the nucleus is assumed to be a sphere of radius r. The energy change of (5.8) is therefore

$$\Delta G_{L\rightarrow S}^{hom} = \frac{\Delta g_{L\rightarrow S}^{nuc}}{\Omega} \cdot \frac{4}{3}\pi r^3 + \gamma_{LS} \cdot 4\pi r^2 \tag{5.9}$$

where γ_{LS} is the energy per unit area of the interface between the liquid and the solid. A nucleus will grow if the energy of the system decreases as the radius expands. Hence the driving force for growth of the particle is

$$F = -\frac{\partial \Delta G_{L\rightarrow S}^{hom}}{\partial r} = -4\pi r^2 \left[\frac{\Delta g_{L\rightarrow S}^{nuc}}{\Omega} + \frac{2\gamma_{LS}}{r} \right]. \tag{5.10}$$

If $F < 0$ then the nucleus will be unstable and shrink and disappear, i.e. it is not large enough to survive. If $F > 0$ then the nuclei will undergo stable growth, forming a larger precipitate. Note that when $F = 0$, the radius of the nucleus is

$$r^* = -\frac{2\gamma_{LS}\Omega}{\Delta g_{L\rightarrow S}^{nuc}}. \tag{5.11}$$

This is known as the critical radius. To be stable, any nucleus that forms must be at or above this size. When $\Delta g_{L\rightarrow S}^{nuc}>0$ the radius is negative and stable nucleation is not possible. At the LIQUID–SOLID phase boundary $\Delta g_{L\rightarrow S}^{nuc} = 0$ and a nucleus would have to be infinite in size to be stable. As the undercooling increases, $\Delta g_{L\rightarrow S}^{nuc}$ becomes more negative and the critical radius for nucleation decreases. The probability of a nucleus randomly forming depends on its size and the kinetics of the material transport required for this to happen. This is considered further in section 7.1.

Substituting (5.11) into (5.10) gives the force acting on the interface to be

$$F = 4\pi r^2 \cdot 2\gamma_{LS}\left[\frac{1}{r^*} - \frac{1}{r}\right] \tag{5.12}$$

which shows more clearly that precipitates with $r > r^*$ grow and those with $r < r^*$ shrink. The activation energy for homogenous nucleation is found by substituting (5.11) into (5.9) to give

$$\Delta G^*_{\text{hom}} = \frac{4}{3}\pi\gamma_{LS}(r^*)^2. \tag{5.13}$$

This is the total change in energy (in Joules) when a nucleus of radius r^* is created. This is positive indicating that there is an energy barrier to nucleation.

The relative size and dependency of these parameters are explored for the fictitious example of section 5.2, illustrated in figure 5.2. The driving force for nucleation for an alloy with $x_0 = 0.2$ is considered as the temperature is lowered from 750 °C to 650 °C. The driving force for nucleation $\Delta g^{\text{nuc}}_{L \to S}$ is plotted in figure 5.4(a). It is well represented by a straight line of the form $\Delta g^{\text{nuc}}_{L \to S} = \Delta S(T - T^M)$ where $T^M = 722$ °C is the melting temperature and $\Delta S = 20.9$ J mol^{-1}. This is almost identical to the entropy change $\Delta S_A = \Delta S_B = 20$ J/mol for this system, as defined in equation (3.12). This is because, in this model, the two values are the same and constant (independent of composition and temperature) and the excess energy is also not temperature dependent. The small difference from the expression for a pure metal, equation (5.1), is due to the entropic terms present in a two phase system. Figure 5.4(b) shows the critical radius r^* for a molar volume of $\Omega = 10^{-5}$ m^3/mol and an interfacial energy of $\gamma_{LS} = 0.02$ J/m^2. The radius is negative above the melting temperature, indicating nucleation is not energetically favourable. By 700 °C, the undercooling is sufficient to reduce the radius to a feasible size for a nucleus of less than 1–2 nanometres. Figure 5.4(c) shows the critical energy barrier for the formation of a nucleus, ΔG^*_{hom}, drops rapidly with temperature below the melting point. At nucleation it is of the order of a few electron volts (1 eV $= 1.6 \times 10^{-19}$ J).

5.3.2 Heterogeneous nucleation

Homogeneous nucleation is rare in practice as inhomogeneities in the system provide more favourable locations for nucleation. For example, the formation of ice from sub-zero water is one of the most commonly observed liquid-to-solid phase transitions. In highly purified water, an undercooling of 35 °C is required to form ice crystals, whereas water typically freezes with an undercooling of 5 °C or less due to the impurities it contains. Nucleation on a surface or some sort of defect is known as **heterogeneous nucleation** and is energetically more favourable than homogeneous nucleation. Typical examples can be observed in every day life. Carbon dioxide

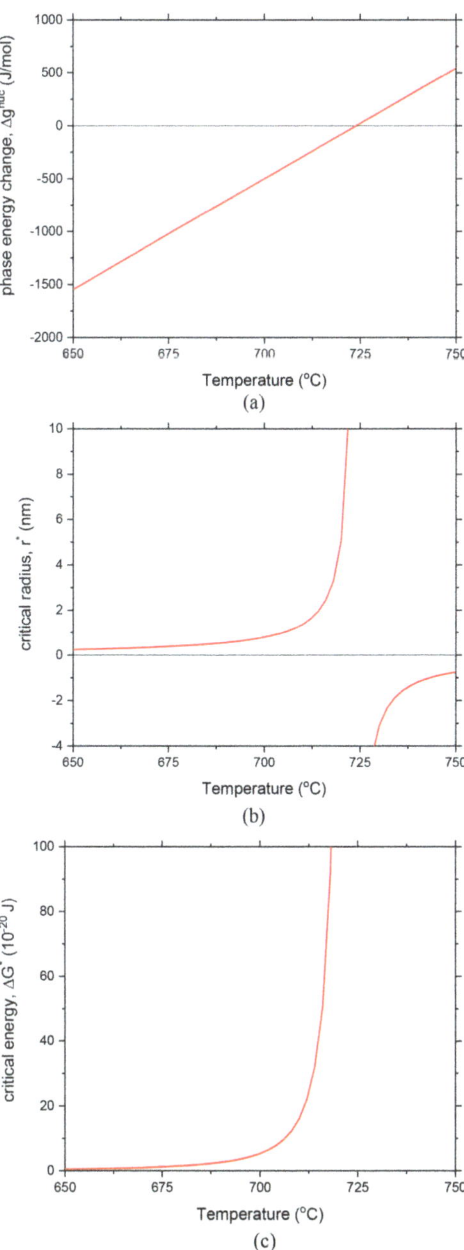

Figure 5.4. Nucleation of the SOLID phase with the LIQUID phase for the alloy of figure 5.2 with $x_0 = 0.2$. (a) the phase energy change calculated from PANDAT, (b) the critical radius calculated from equation (5.11) with $\Omega = 10^{-5}$ m^3 mol^{-1} and $\gamma_{LS} = 0.02$ J m^{-2}, (c) the critical energy barrier ΔG^*_{hom} of equation (5.13).

bubbles in a fizzy drink typically nucleate on the surface of the glass, where small grooves or dirt particles assist the process. Adding a sugar cube to a carbonated drink provides many more nucleation sites, causing it to erupt with froth due to the rapid release of carbon dioxide. Water vapour in the exhaust gases from aircraft jets engines nucleates on impurities in the fuel. The resulting droplets can freeze into ice crystals at high altitudes forming the contrails that we can see from the ground.

5.3.2.1 Nucleation on a free surface

A common example of this is the gas-to-liquid phase transition seen as condensation of water droplets onto a cold surface, such as a window or a bathroom mirror. A spherical cap is used to represent the shape of the liquid droplet on the free surface, as shown in figure 5.5(a). This same analysis applies to the formation of a solid nucleus from the liquid melt, perhaps on the surface of a casting mould or an existing particle in the melt. Three interfacial energies now need to be considered: γ_{FL} the energy per unit area between the free surface (denoted by F) and the surrounding liquid; γ_{FS} the energy between the solid droplet and the surface; and γ_{LS}, the energy between the liquid and the solid nucleus. The geometry of the system is defined by the radius of curvature of the droplet, r, and the contact angle ψ. The contact angle must satisfy force balance at the triple point, as shown in figure 5.5(a), such that

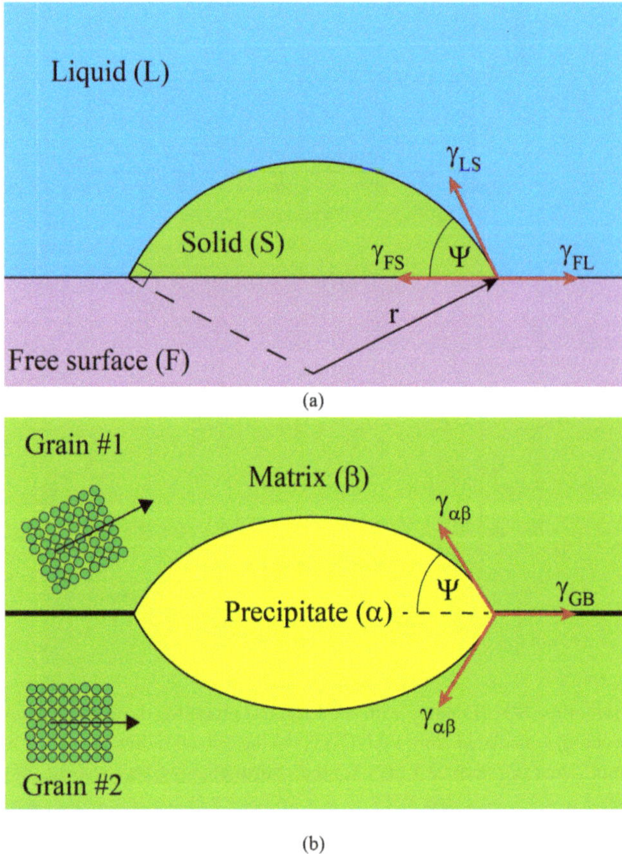

Figure 5.5. Heterogeneous nucleation of a solid particle on (a) from a liquid onto a free surface, and (b) from a solid matrix phase onto a grain boundary.

$$\gamma_{LS}\cos\psi + \gamma_{FS} = \gamma_{FL}. \tag{5.14}$$

Following (5.9), the change in energy of the system upon heterogeneous nucleation of the droplet is

$$\Delta G_{L\to S}^{het} = \frac{\Delta g_{L\to S}^{nuc}}{\Omega} \cdot V_S + \gamma_{LS}A_{LS} + (\gamma_{FS} - \gamma_{FL})A_{FS} \tag{5.15}$$

where a liquid–solid interface is created at the free surface area of the solid nucleus $A_{LS} = 2\pi r^2(1 - \cos\psi)$, and the liquid-surface interface is replaced by a solid-surface interface at the area of contact between them, $A_{FS} = \Pi r^2(1 - \cos^2\psi)$. The volume of the spherical cap is $V_S = \frac{2}{3}\pi r^3 \cdot K(\psi)$, where

$$K = \frac{1}{2}(2 - 3\cos\psi + \cos^3\psi). \tag{5.16}$$

Substitution of equation (5.14) into (5.15) gives

$$\Delta G_{L\to S}^{het} = \frac{1}{2}K(\psi)\Delta G_{L\to S}^{hom}. \tag{5.17}$$

As the heterogeneous energy change simply differs by a geometric factor from the homogeneous energy change, the critical radius r^* is still defined by equation (5.11), although it should be noted that this is now just the radius of curvature of the droplet and that the critical nucleus size is much reduced. The critical nucleation energy is also reduced

$$\Delta G_{het}^{*} = \frac{1}{2}K(\psi)\Delta G_{hom}^{*} \tag{5.18}$$

where ΔG_{hom}^{*} is defined by equation (5.13). The function $K(\psi)$ is plotted in figure 5.6 and shows that the energy barrier for nucleation can be greatly reduced on a free surface, especially when the contact angle is very low ($<30°$). As expected, the homogeneous result is recovered when the contact angle is $180°$, i.e. there is no contact with the free surface.

5.3.2.2 Nucleation on a grain boundary

For solid-to-solid phase transitions, different preferential nucleation sites are typically defects such as grain boundaries (GBs), dislocations, vacancies etc with higher energy than the uniform matrix. The formation of nuclei on these sites is accompanied by a reduction in energy due to (partial) removal of the defect and hence is energetically more favourable. A double spherical cap model is applied for the precipitation of a solid phase β on a grain boundary between two crystals of phase α. The interfacial energy between the two phases is $\gamma_{\alpha\beta}$ and the interfacial energy of the grain boundary is γ_{GB}, as shown in figure 5.5(b). The contact angle at the triple point is defined by interfacial tension balance

$$2\gamma_{\alpha\beta}\cos\psi = \gamma_{GB}.$$

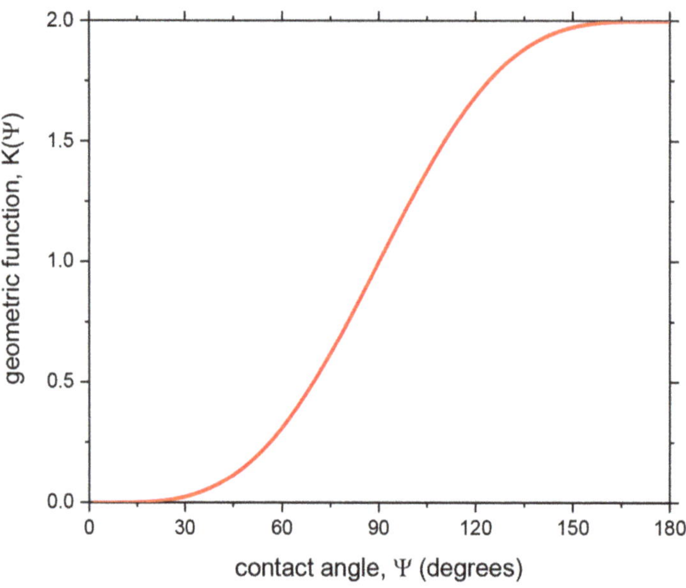

Figure 5.6. The geometric function $K(\psi)$ of equation (5.16).

This is similar to equation (5.14) with $2\gamma_{\alpha\beta} = \gamma_{\text{LS}}$, $\gamma_{\text{GB}} = \gamma_{\text{FL}}$ and $\gamma_{\text{FS}} = 0$. Hence the same conclusions from the free surface nucleation case can be drawn, but now the volume and interfacial area of the nucleus are multiplied by two due to the double spherical cap geometry of the precipitate. The critical radius of curvature r^* is therefore still defined by equation (5.11) and the critical energy for heterogeneous nucleation on a grain boundary is double that for the free surface case in equation (5.18)

$$\Delta G^{*}_{\text{het (GB)}} = K(\psi)\Delta G^{*}_{\text{hom}}. \tag{5.19}$$

The solid-to-solid phase energy change $\Delta g^{\text{nuc}}_{\alpha\rightarrow\beta}$ is found for the two relevant phases using equation (5.4). When the grain boundary is removed ($\gamma_{\text{GB}} = 0$) the contact angle $\psi = 90°$ such that $K = 1$ and the result for homogeneous nucleation is recovered as expected.

Reference

[1] Pandat: software suite for thermodynamic calculation and kinetic simulation of multi-component alloys. CompuTherm LLC, Madison, Wisconsin, USA www.computherm.com [accessed 1 October 2021]

Part II

Kinetics and microstructure evolution

IOP Publishing

Thermodynamics, Kinetics and Microstructure Modelling

Simon P A Gill

Chapter 6

Kinetic processes

The first half of this book introduced the concept of thermodynamics and equilibrium phase diagrams. We learnt that the Gibbs free energy provides the driving force for phase transformations, but how phases form or how long this might take was not considered. If a material is to evolve over time then there must be a kinetic process by which it can do this. This is the focus of the second half of this book, where models for the movement of atoms are introduced. Different phases form over time by the movement of atoms through kinetic processes such as diffusion. Diffusion increases rapidly with temperature, and hence materials change more rapidly as the temperature is increased. As such, a material system may not always reach its equilibrium state, either because the driving force for change is weak, or the kinetics are too slow at the given temperature. As seen in chapter 5, additional factors such as the interfacial energy between phases can also be influential.

In reality, some equilibrium phases do not appear until a material has been thermally aged for some time, in some cases for many years. Non-equilibrium phases, that do not appear on the phase diagram, may exist in the meantime if they nucleate or grow faster. This slow evolution to equilibrium can be beneficial: high strength steels can be formed by rapid cooling, where the matrix is trapped in its non-equilibrium martensitic phase at lower temperatures. It can also be problematic: if the equilibrium phases that appear very late are deleterious to the performance of the material, then these may cause a material to unexpectedly degrade after many years of service. Stainless steels used in steam turbines are an example of this. They are under constant stress at high temperature for the lifetime of the plant, typically 25 years or more. It is known that a number of brittle phases start to appear after 5–10 years under these conditions. These phases may not appear in the duration of the experimental tests that are used to validate the material, as these are practically time limited to 1 or 2 years. Therefore it is important to be able to predict what will happen to a material over long periods of time through modelling.

In addition, the distribution of the phases that form is very important. The ability of precipitates to act as obstacles to dislocation motion depends very much on their size, as described in section 1.2. Generally it is much more beneficial to have lots of very small precipitates rather than a few larger ones. Many small particles appear if nucleation is fast but then their growth is slow. A smaller number of larger particles results if nucleation is slow, but their growth is then fast. Both **nucleation** and **growth** are controlled by both the thermodynamics and kinetics of the system. But the evolution of a material does not stop there, as it will wish to reduce its interfacial energy over time. This process, known as **coarsening**, occurs more slowly and is generally detrimental as it results in an increase in precipitate size and a decrease in precipitate number. These three processes are linked and are considered together in chapter 7.

The introduction of kinetics also means the introduction of space and time into the model. Following the common nomenclature in kinetic modelling, the mol% concentrations of elements will be denoted by the variable c rather than the x adopted by the thermodynamics community and used in the first half of this book. The variable x will now be used to denote spatial position.

6.1 An atomic model of diffusion

Diffusion is the process by which atoms move within a crystal. This occurs by an atom relocating to a vacant neighbouring site in the lattice. There are two types of diffusive process: interstitial and substitutional. **Interstitial diffusion** is when an interstitial atom hops from one interstitial site to a vacant neighbouring interstitial site, as shown in figure 6.1. Similarly, **substitutional diffusion** is from one substitutional site to another,

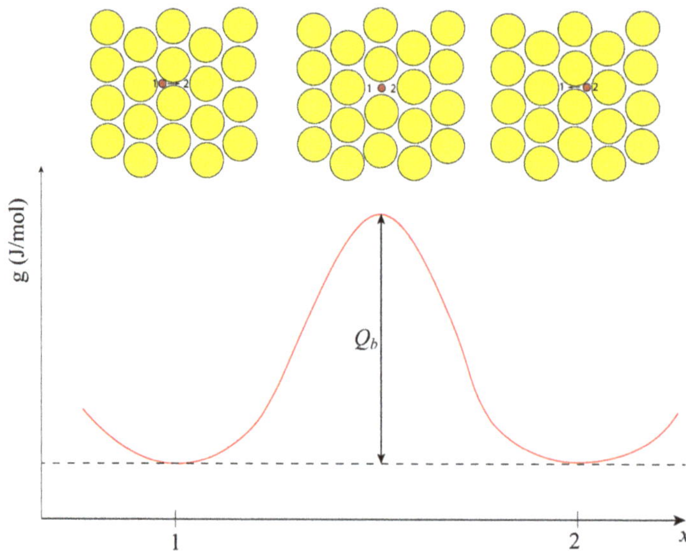

Figure 6.1. A schematic diagram showing three snapshots of an atom moving between substitutional lattice sites 1 and 2 in a unary crystal. The energy of the system increases by an amount Q_b as the atom squeezes between its neighbours due to local distortion of the lattice, limiting the rate of the process.

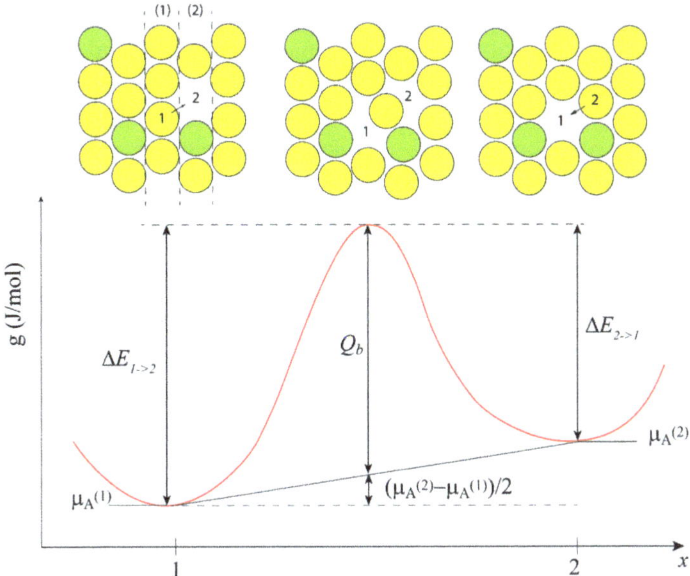

Figure 6.2. Substitutional diffusion is mediated by vacancies. A schematic of the chemical potential change during substitutional diffusion in a multi-component alloy of species A (yellow) and B (green). In this case, the example shows an atom moving between two substitutional lattice sites, labelled 1 and 2. The chemical potential of site 2 is higher than that of site 1. This introduces an asymmetry into the energy profile, reducing the energy barrier for a hop from site 2 to 1, $\Delta E_{2\rightarrow 1}$, and increasing the energy barrier for a hop from site 1 to 2, $\Delta E_{1\rightarrow 2}$, resulting in a net flow of atoms from right to left, down the chemical potential gradient. The chemical potential change is calculated by considering the change in energies of the atoms in slice (1) and slice (2) before and after the transition.

as shown in figure 6.2. The main difference between the two is that most interstitial sites are typically empty, whereas most substitutional sites are not. Therefore substitutional diffusion is strongly limited by the population of vacancies. Interstitial diffusion is consequently much faster than substitutional diffusion.

As illustrated in figure 6.1, the atomic lattice must locally distort during the diffusive process to allow an atom to pass between minimum energy sites. This temporarily raises the energy (and chemical potential) of the system by an amount q_b (in Joules) or by $Q_b = N_{Av}q_b$ in Joules per mole, where N_{Av} is Avagradros constant for the number of atoms in a mole. The quantity Q_b is known as the **activation energy** or energy barrier to diffusion. This controls the rate of the hopping process, with a larger Q_b leading to a lower hopping rate.

If there is no net change in the internal energy after a hop is completed, as shown in figure 6.1, then the diffusion process occurs randomly, and the net flow of atoms in one direction is equal to the flow in the opposite direction. In a single component system, this results in no net change to the system. However, in a multi-component system, this random hopping of atoms back and forth will cause the different elements to mix. At a continuum level, the driving force for this mixing is represented by the contribution of entropy, considered in chapter 3, which drives a system towards its most probable (fully mixed) state.

A thermodynamic driving force for diffusion will introduce a bias in the hopping process, leading to more atomic jumps of a particular species in one direction than another. This generates a net flux of atoms in that direction. Consider an atom of species A hopping between lattice sites 1 and 2, as shown in figure 6.2. The probability of a hop to the right ($1 \rightarrow 2$) being successful is

$$P_{1 \rightarrow 2} = \exp\left(-\frac{\Delta e_{1 \rightarrow 2}}{k_B T}\right) = \exp\left(-\frac{\Delta E_{1 \rightarrow 2}}{RT}\right)$$

where $\Delta e_{1 \rightarrow 2}$ and $k_B T$ are the potential energy barrier associated with a hop from 1 to 2 and the kinetic energy of this degree-of-freedom of the atom in Joules. The parameters $\Delta E_{1 \rightarrow 2} = N_{av}\Delta e_{1 \rightarrow 2}$ and $R = N_{av}k_B$ are the equivalent quantities in Joules per mole. A similar expression applies for the probability of a successful reverse jump, $P_{2 \rightarrow 1}$. Note that if there is no energy barrier to a transition then the probability of a successful jump is 1, i.e. every attempt is successful.

Following equation (3.7), the energy of the donor slice (1) in Joules is $g^{(1)} = [n_A^{(1)}\hat{\mu}_A^{(1)} + n_B^{(1)}\hat{\mu}_B^{(1)}]$ and that of the receptor slice (2) is $g^{(2)} = [n_A^{(2)}\hat{\mu}_A^{(2)} + n_B^{(2)}\hat{\mu}_B^{(2)}]$, where $\hat{\mu}_A^{(1)} = \dfrac{\mu_A^{(1)}}{N_{av}}$ is the average chemical potential of A per atom in slice (1) and $n_A^{(1)}$ is the number of A atoms in slice (1) etc. Now after the diffusion event the A atom has left slice (1), so $n_A^{(1)} \rightarrow n_A^{(1)} - 1$, and entered slice (2), such that $n_A^{(2)} \rightarrow n_A^{(2)} + 1$. The total energy of the two slices is initially $g^{(1)} + g^{(2)}$. After the diffusion event the total energy is $g^{(1)} + g^{(2)} + \hat{\mu}_A^{(2)} - \hat{\mu}_A^{(1)}$. The change in the Gibbs free energy due to the exchange is therefore due to a change in chemical potential of $\hat{\mu}_A^{(2)} - \hat{\mu}_A^{(1)}$ Joules for one diffusing atom, or $\mu_A^{(2)} - \mu_A^{(1)}$ J mol^{-1} for a mole of diffusing atoms.

The change in the energy transition profile due to an atomic transition such as this is shown in figure 6.2, for the case where $\mu_A^{(2)} > \mu_A^{(1)}$. The potential barrier for a transition is modified by the chemical potential gradient. In this case, the activation energy for a hop from 1 to 2 is slightly increased, $\Delta E_{1 \rightarrow 2} = Q_b + \dfrac{1}{2}\Delta\mu_A$, and the activation energy for a hop from 2 to 1 is slightly decreased, $\Delta E_{2 \rightarrow 1} = Q_b - \dfrac{1}{2}\Delta\mu_A$, where $\Delta\mu_A = \mu_A^{(2)} - \mu_A^{(1)}$. It is therefore expected that an atomic transition from site 2 to 1 is more favourable than a transition from 1 to 2. The average velocity of atoms of A from site 1 to 2 depends on the difference between the two hopping rates

$$v_A = a\nu c_v c_A [P_{1 \rightarrow 2} - P_{2 \rightarrow 1}] = -a\nu c_v c_A \exp\left(-\frac{Q_b}{RT}\right)\sinh\left(\frac{\Delta\mu_A}{2RT}\right) \qquad (6.1)$$

where a is the jump distance (roughly equivalent to the atomic spacing), ν is the attempt frequency (the vibrational frequency of the atoms), c_A is the probability that the donor site is occupied by an atom of species A (the concentration of species A) and c_v is the probability that the receptor site is vacant (the concentration of vacancies). As expected, this predicts that there is a net flow from a higher chemical potential to a lower chemical potential, i.e. from site 2 to 1 in the example of figure 6.2.

The majority of interstitial sites are vacant so $c_v \approx 1$ for interstitial diffusion. For substitutional diffusion, the probability of a lattice site being empty is

$$c_v = \exp\left(-\frac{Q_v}{RT}\right)$$

where Q_v (J mol^{-1}) is the energy required to create a mole of vacancies. If $\Delta\mu \ll RT$ then it is reasonable to write (6.1) as

$$v_A = -a v c_A \exp\left(-\frac{Q_A}{RT}\right)\frac{\Delta\mu}{2RT} \tag{6.2}$$

where the activation energy for a substitutional species, $Q_A = Q_b + Q_v$, is typically much higher than for an interstitial species, $Q_A = Q_b$. As such, interstitial diffusion is usually much faster than substitutional diffusion.

Within a phase the chemical potential variation will be continuous such that $\Delta\mu \approx \frac{\partial\mu}{\partial x}a$. Material transport is typically defined in terms of the flux. The flux of species A is represented by j_A. This is the volume of A atoms crossing a unit area per second, and has units of metres per second. Assuming that the volume of an atom of A is a^3 and it jumps across a cross-sectional area a^2 then the flux is equivalent to the net velocity, i.e. $j_A = v_A$. Hence the flux of A through a phase p can be determined from (6.2) to be

$$j_A = -\frac{c_A D_A^p}{RT}\frac{\partial\mu_A^p}{\partial x} \tag{6.3}$$

where $D_A^p = \frac{1}{2n}a^2 v \exp\left(-\frac{Q_A^p}{RT}\right)$ is known as the **self-diffusivity** (or tracer diffusivity) of A in phase p and has units of m^2 s^{-1}, and $n = 1, 2$ or 3 is introduced to represent the effect of diffusion in one-, two- or three-dimensions. The temperature dependence of the diffusivity is Arrhenius, showing that the diffusion of atoms increases with temperature. This constitutive law shows that atoms diffuse down a chemical potential gradient (due to the minus sign) in order to reduce the overall Gibbs free energy of the system. It was shown in section 3.2.2 that the chemical potential of a component is the same at all points within a system in its minimum (equilibrium) energy state. In this state there is no chemical potential gradient and hence, from equation (6.3), there is no material transport, i.e. the state does not change, as expected.

Equation (6.3) is also sometimes written as

$$j_A = -c_A M_A^p \frac{\partial\mu_A^p}{\partial x} \tag{6.4}$$

where $M_A^p = \frac{D_A^p}{RT}$ is known as the **mobility** of A in phase p.

Diffusion through a crystal lattice is easier when the atoms are not so closely packed. Regions of lower packing density are generally associated with crystallographic defects, as discussed in section 1.2.3. As such, enhanced diffusivities are associated with material transport along continuous defects such as dislocation cores (1D) and grain boundaries (2D).

6.2 Defining diffusion parameters in TDB files

TDB files were encountered in Part I to define the thermodynamics of different phases in the form of the Gibbs free energy. TDB files can also be used to store kinetic data.

6.2.1 Diffusion parameters in a unary system

Consider the definition of diffusivity in the Al unary system. The solid phase in this system is FCC, and hence only one diffusion coefficient needs to be defined: the self-diffusivity of Al in the pure Al FCC phase, written as $D_{\mathrm{Al}}^{\mathrm{FCC(Al)}}$, or the associated mobility $M_{\mathrm{Al}}^{\mathrm{FCC(Al)}}$. The mobility of species A in phase p can be written in general form as $M_{\mathrm{A}}^{\mathrm{p}} = M_{\mathrm{A0}}^{\mathrm{p}} \exp\left(-\dfrac{Q_{\mathrm{A}}^{\mathrm{p}}}{RT}\right)$. Its value can change by many orders of magnitude over the relevant temperature range. As such it is defined by the parameter MQ, where

$$\mathrm{MQ} = RT \ln\left(M_{\mathrm{A}}^{\mathrm{p}}\right) = RT \ln\left(M_{\mathrm{A0}}^{\mathrm{p}}\right) - Q_{\mathrm{A}}^{\mathrm{p}}. \qquad (6.5)$$

The parameter $M_{\mathrm{Al}}^{\mathrm{FCC(Al)}}$ is defined for diffusion of Al in FCC Al in the following TDB code. The first 3 lines simply define the element Al and the solid phase FCC, as seen before. The last line defines the mobility of Al in pure FCC Al matrix, setting $M_{\mathrm{Al0}}^{\mathrm{FCC(Al)}} = 0.000171$ and $Q_{\mathrm{Al}}^{\mathrm{FCC(Al)}} = 142$ kJ mol^{-1}.

Defining unary diffusion data in a TDB file #1: MQAl.tdb

```
ELEMENT AL  FCC_A1  26.982  4540  28.3 !

PHASE FCC_A1%  1  1 !
CONSTITUENT FCC_A1  :AL: !
PARAMETER   MQ(FCC_A1&AL,AL;0)    298.15    -142000+R*T*LN
(0.000171); 6000 N !
```

The tracer diffusivity (or self-diffusivity) can be defined using the parameter DT, and the chemical diffusivity by the parameter DC (see equation (6.7)). The following instructions show how PANDAT can be used to plot the variation in the mobility of Al in FCC Al over the temperature range of 0° to 660 °C.

PANDAT: plotting diffusion parameters using 'MQAl.tdb'

1. Copy and paste the contents of 'MQAl.tdb' above into a text editor and save it into a file of that name.
2. Run Pandat. Under 'Create a New Workspace', select the 'PanPhaseDiagram' icon and click the 'Create' button.
3. Select 'Databases → Load TDB or PDB (Encrypted TDB)' and open the file 'MQAl.tdb'.
4. Under 'Select Components', click on the 'Sel/Clr All' button to move all the 'Available Components' into the 'Selected Components' column. Press 'OK'.
5. Go to 'Property → Kinetic Property'.
6. Set the start point temperature 'T(C)' to '0', and the end point temperature to 'T(C)' to '660'.
7. Note that in the 'Choose Target Kinetic Properties for Plot' the checkbox is checked next to the mobility 'M(*@*)'.
8. Under 'Condition for Chemical Diffusivity' set the 'Reference Species(N)' to 'Al'. Click OK.
9. The mobility is shown in figure 6.3(a). It appears to be zero below 400 °C, but in fact it is just very small. To see this, go back to 'Property → Kinetic Property', uncheck the checkbox next to 'M(*@*)' and check the 'log10(M(*@*))' checkbox.
10. The exponent $\log_{10}(M)$ is plotted in figure 6.3(b). This shows that the mobility of Al changes by 18 orders of magnitude over the 660 °C temperature range!

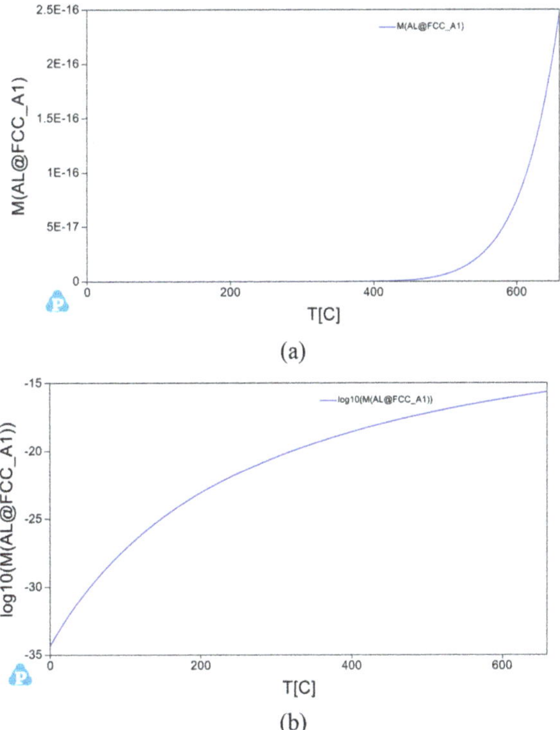

(a)

(b)

Figure 6.3. The mobility of Al in an FCC Al matrix as a function of temperature for (a) the mobility $M_{\mathrm{Al}}^{\mathrm{FCC(Al)}}$ and (b) $\log_{10}(M_{\mathrm{Al}}^{\mathrm{FCC(Al)}})$. Created using PANDAT [8].

6.2.2 Diffusion parameters in a binary system

For a binary system containing species A and B, the mobility of both species need to be defined as a function of the composition of the matrix phase. This takes a similar polynomial form to that used to define the Gibbs free energy, such that the mobilities of A and B in an alloy matrix of phase p and composition c_A and c_B is given by

$$M_A^{p(AB)}(c_A, c_B) = M_A^{p(A)}c_A + M_A^{p(B)}c_B + N_A^{p(AB)}c_A c_B$$

$$M_B^{p(AB)}(c_A, c_B) = M_B^{p(A)}c_A + M_B^{p(B)}c_B + N_B^{p(AB)}c_A c_B$$

where $M_A^{p(A)}$ is the mobility of A in phase p of pure A, $M_A^{p(B)}$ is the mobility of A in phase p of pure B etc. The kinetic interaction parameters, $N_A^{p(AB)}$ and $N_B^{p(AB)}$, can themselves be composition dependent, and are given in the Redlich–Lister form of equation (3.6). PANDAT provide the following example mobilities for the Ni–Al binary system. The last six lines define $M_{Al}^{FCC(Al)}$ (as in the previous section), $M_{Al}^{FCC(Ni)}$, $N_{Al}^{FCC(AlNi)(0)}$ and $M_{Ni}^{FCC(Al)}$, $M_{Ni}^{FCC(Ni)}$, $N_{Ni}^{FCC(AlNi)(0)}$ in the format of equation (6.5) respectively.

Defining binary diffusion data in a TDB file #2: MQNiAl.tdb

```
ELEMENT AL FCC_A1 26.982 4540 28.3 !
ELEMENT NI FCC_A1 58.693 4787 29.796 !


PHASE FCC_A1% 1 1 !
CONSTITUENT FCC_A1 :AL,NI:!
PARAMETER     MQ(FCC_A1&AL,AL;0)     298.15-142000+R*T*LN
(0.000171); 6000 N !
PARAMETER     MQ(FCC_A1&AL,NI;0)     298.15-284000+R*T*LN
(0.00075); 6000 N !
PARAMETER    MQ(FCC_A1&AL,AL,NI;0)   298.15-41300-91.2*T;
6000 N !
PARAMETER     MQ(FCC_A1&NI,AL;0)     298.15-145900+R*T*LN
(0.00044); 6000 N !
PARAMETER MQ(FCC_A1&NI,NI;0) 298.15-287000-69.8*T; 6000
N !
PARAMETER    MQ(FCC_A1&NI,AL,NI;0)   298.15-113000+65.5*T;
6000 N !
```

The following instructions show how PANDAT can be used to plot the variation in the mobility of Al and Ni in the FCC phase of a NiAl at a temperature of 600 °C as a function of the matrix composition.

PANDAT: plotting diffusion parameters using 'MQNiAl.tdb'

1. Copy and paste the contents of 'MQNiAl.tdb' above into a text editor and save it into a file of that name.
2. Run Pandat. Under 'Create a New Workspace', select the 'PanPhaseDiagram' icon and click the 'Create' button.
3. Select 'Databases → Load TDB or PDB (Encrypted TDB)' and open the file 'MQNiAl.tdb'.
4. Under 'Select Components', click on the 'Sel/Clr All' button to move all the 'Available Components' into the 'Selected Components' column. Press 'OK'.
5. Go to 'Property → Kinetic Property'.
6. Set the start point temperature 'T(C)' to '600' and the composition 'x(Al)' to '1' and 'x(Ni)' to '0'. Set the end point temperature to 'T(C)' to '600' and the composition 'x(Al)' to '0' and 'x(Ni)' to '1'
7. Note that in the 'Choose Target Kinetic Properties for Plot' the checkbox is checked next to the mobility 'M(*@*)'.

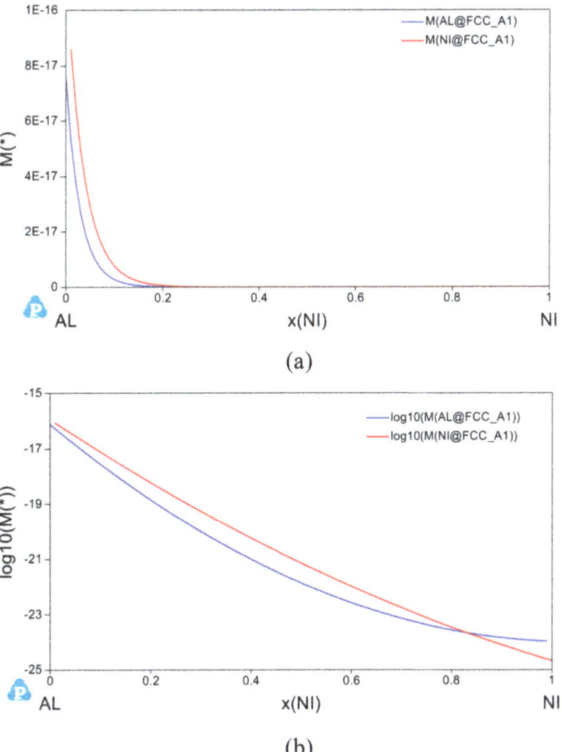

Figure 6.4. The mobility of Al and Ni in an FCC Al/Ni alloy matrix at 600 °C as a function of its composition for (a) the mobilities $M_{Al}^{FCC(AlNi)}$ and $M_{Ni}^{FCC(AlNi)}$ and (b) $\log_{10}(M_{Al}^{FCC(AlNi)})$ and $\log_{10}(M_{Ni}^{FCC(AlNi)})$. Created using PANDAT [8].

8. Under 'Condition for Chemical Diffusivity' set the 'Reference Species(N)' to 'Al'. Click OK. Click 'OK'.
9. The results are shown in figure 6.4(a). It is clear that the mobility of both species is very much higher in an Al rich matrix than a Ni rich matrix. This is largely because the temperature is much nearer the melting point of Al (660 °C) than that of Ni (1455 °C).
10. The lower range of diffusivities is seen more clearly by changing the plot parameter from 'M(*@*)' to 'log10(M(*@*))' as described in the previous subsection. The results are plotted in figure 6.4(b). The diffusivities are similar for Al and Ni, suggesting they are strongly controlled by the matrix they exist within. They vary by 8–9 orders of magnitude between pure Al and pure Ni.

6.3 Diffusion in a regular solution

The chemical potential for the regular solution model is determined in equation (3.9) as

$$\mu_A^p(c_A) = g_p^A + L_p(1 - c_A)^2 + RT \ln(c_A). \tag{6.6}$$

The chemical potential, introduced in section 3.1.6, is a function of composition, such that equation (6.3) can be expressed as

$$j_A = -\frac{c_A D_A^p}{RT} \frac{\partial \mu_A^p}{\partial c_A} \cdot \frac{\partial c_A}{\partial x} = -D_A^{p*} \frac{\partial c_A}{\partial x} \tag{6.7}$$

where $D_A^{p*} = \dfrac{c_A D_A^p}{RT} \dfrac{\partial \mu_A^p}{\partial c_A}$ is known as the **chemical diffusivity** of A in phase p. Substitution of equation (6.6) yields

$$D_A^{p*} = D_A^p \left[1 - 2c_A(1 - c_A)\frac{L_p}{RT} \right]. \tag{6.8}$$

The first term within the brackets (the number 1) is due to the entropic contribution and always ensures a positive contribution to the diffusion coefficient. This drives atoms down a concentration gradient, causing species to mix. However, the second (L_P) term can cause the overall diffusivity to be negative, leading to species separating (unmixing). If $2c_A(1 - c_A)\dfrac{L_p}{RT} > 1$ then $D_A^{p*} < 0$ and the excess energy drives diffusion up the concentration gradient. This is known as **uphill diffusion**. Recall that, in figure 3.3, a single phase binary material is shown to decompose into A rich and B rich regions in the SOLID + SOLID region of the phase diagram. This is due to the excess energy term and is a consequence of uphill diffusion.

The first model for diffusion was proposed by Fick in 1855. This assumes **purely entropic diffusion**, neglecting the contribution due to the excess energy $(L_p = 0)$, giving $D_A^{p*} = D_A^p$ and **Fick's first law**

$$j_A = -D_A^p \frac{\partial c_A}{\partial x} \tag{6.9}$$

in which $D_A^p > 0$ and hence diffusive transport is always down a concentration gradient. Equation (6.8) shows that Fickian diffusion is particularly pertinent for dilute solutes, i.e. $c_A \ll 1$.

In higher dimensions, the flux is a vector. The three-dimensional form of equation (6.7) is

$$\underline{j}_A = \left[j_x^A, j_y^A, j_z^A \right] = -D_A^{p*} \left[\frac{\partial c_A}{\partial x}, \frac{\partial c_A}{\partial y}, \frac{\partial c_A}{\partial z} \right]. \tag{6.10}$$

For multi-component diffusion, the chemical potential is a function of multiple independent components, i.e. $\mu_A^P(c_A, c_B, c_C, ...)$. In this case, using the chain rule on equation (6.7), coupled with equation (6.10), the flux vector becomes

$$\underline{j}_A = -\sum_i D_{Ai}^{p*} \left[\frac{\partial c_i}{\partial x}, \frac{\partial c_i}{\partial y}, \frac{\partial c_i}{\partial z} \right] \tag{6.11}$$

where the latter expression is for purely entropic diffusion. In this case gradients in the concentration of any component can drive the diffusion of all the other components and $D_{Ai}^{p*} = \dfrac{c_A D_A^P}{RT} \dfrac{\partial \mu_A^P}{\partial c_i}$ is the chemical diffusivity of A due to gradients in component i. Note that in this general format the parameter D_{AA}^{p*} is equivalent to D_A^{p*}.

Additional complications can also arise due to the choice of reference frame. As all the components of a matrix may be diffusing, there is no fixed reference frame. For this reason, the flux is often taken relative to the concentration-weighted average flux, $\bar{j} = \sum_i c_i j_i$. In this case, the flux of A would be taken to be $j_A - \bar{j}$. This adjustment is not considered further here for the benefit of simplicity.

6.4 Mass conservation

The concentration of species A at a point in a material will change if the inward flux of that species is different to the outward flux. Consider an infinitesimal cube of width dx, as shown in figure 6.5, subject to inward and outward fluxes j_A and $j_A + dj_A$ in the x-direction across the cross-sectional area A_S. The volume of the cube is $A_S dx$. If the concentration of species A in the cube is c_A, then its volume is $c_A A_S dx$. The volumetric flow rate of species into the cube is $j_A A_S$ and the flow rate out of the cube $(j_A + dj_A) A_S$ is slightly bigger or smaller. The rate of change in the volume of species A is the difference between the inward and outward fluxes such that

$$\frac{\partial (c_A A_S dx)}{\partial t} = j_A A_S - \left(j_A + dj_A \right) A_s = -dj_A A_s.$$

Dividing both sides by the constant $A_S dx$ gives the one-dimensional **mass conservation** equation

$$\frac{\partial c_A}{\partial t} = -\frac{\partial j_A}{\partial x}. \tag{6.12}$$

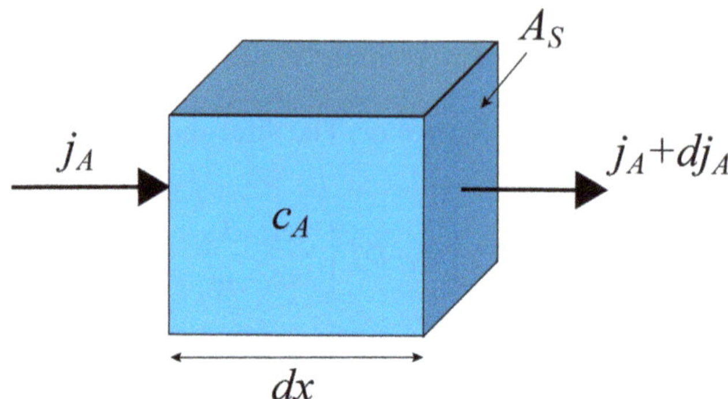

Figure 6.5. Schematic diagram of material flux into and out of a point in a material.

This shows that gradients in the flux cause changes in composition. A negative flux gradient $\left(\dfrac{\partial j_A}{\partial x} < 0\right)$ leaves material behind causing the solute concentration to rise. A positive flux gradient $\left(\dfrac{\partial j_A}{\partial x} > 0\right)$ takes material away causing the solute concentration to drop.

In three-dimensions, equation (6.12) becomes

$$\frac{\partial c_A}{\partial t} = -\left(\frac{\partial j_x^A}{\partial x} + \frac{\partial j_y^A}{\partial y} + \frac{\partial j_z^A}{\partial z} \right). \tag{6.13}$$

6.5 Determining the concentration field

Combining equation (6.13) with equation (6.10) gives

$$\frac{\partial c_A}{\partial t} = \frac{\partial}{\partial x}\left(D_A^{p*}\frac{\partial c_A}{\partial x} \right) + \frac{\partial}{\partial y}\left(D_A^{p*}\frac{\partial c_A}{\partial y} \right) + \frac{\partial}{\partial z}\left(D_A^{p*}\frac{\partial c_A}{\partial z} \right) = \nabla \cdot (D_A^{p*}\nabla c_A). \tag{6.14}$$

In practice the chemical diffusivity is a strong function of composition and temperature. However, as seen in equation (6.9), for a purely entropic system, $D_A^{p*} = D_A^P$. If the system is also isothermal then it is reasonable to assume that D_A^P is a constant, yielding **Fick's second law**

$$\frac{\partial c_A}{\partial t} = D_A^P\left(\frac{\partial^2 c_A}{\partial x^2} + \frac{\partial^2 c_A}{\partial y^2} + \frac{\partial^2 c_A}{\partial z^2} \right) = D_A^P\nabla^2 c_A \tag{6.15}$$

where ∇^2 is the Laplacian operator.

The advantage of equation (6.15) is that there are some known analytical solutions for simple cases [1, 2]. However, in general, it is necessary to solve either equation (6.14) or equation (6.15) numerically. The two most widely used methods are the finite difference method [3] and the finite element method [4, 5]. (Note that equation (6.15) has an identical form to the heat conduction equation, and as such they share the same solutions and solution methods.) To calculate the full concentration field as a function of space and time, $c_A(\underline{x}, t)$, it is necessary to properly formulate the diffusion problem by the specification of the initial conditions and boundary conditions. The initial conditions, $c(x,0)$, define how the concentration of the species is spatially distributed at the start of the simulation, $t = 0$. There are three types of common boundary conditions:

(i) *Dirichlet boundary condition*

Here a quantity of interest is fixed at the boundary. In the purely entropic case of equation (6.15) this would be a prescribed composition at the boundary, where $\underline{x} = \underline{x}^*$, such that

$$c_A(\underline{x}^*, t) = c_0^*(\underline{x}^*) \tag{6.16}$$

where an asterisk denotes a boundary quantity. For the more general case it might be more pertinent to specify the chemical potential on the boundary

$$\mu_A(\underline{x}^*, t) = \mu_0^*(\underline{x}^*).$$

(ii) *Neumann boundary condition*

Here the gradient of a quantity is specified at the boundary. In diffusion simulations the flux is often prescribed

$$j_A(\underline{x}^*, t) = j_0^*(\underline{x}^*) \tag{6.17}$$

where equations (6.3) and (6.9) respectively show that this is equivalent to specifying the chemical potential gradient or the concentration gradient. The most commonly used Neumann condition is the zero flux condition, $j_0^*(\underline{x}^*) = 0$. This is used on symmetry boundaries where there is no net flux of material across the boundary.

(iii) *Robin boundary condition*

Here the flux across a boundary depends on the difference in the conditions on either side of it. The boundary is typically imagined to be in contact with an external reservoir of material (in either solid, liquid or gaseous form) which is a constant source or sink of material. Assume that the chemical potential of element A in the external reservoir is μ_{ext} and that it is μ_A^* at the boundary. At an interface the chemical potential difference may be large, i.e. $\left| \mu_{ext} - \mu_A^* \right| > RT$. If this is the case, the most general form of the transition flux defined in equation (6.1) must be used. Here the asymmetry in the transition rates is defined by

$$j^* = a\nu c_{ext}\left[\exp\left(-\frac{Q_b}{RT}\right) - \exp\left(-\frac{Q_b + \mu_{ext} - \mu_A^*}{RT}\right) \right]$$

or

$$j^* = \alpha c^{\mathrm{ext}}\left[1 - \exp\left(-\frac{\mu_{\mathrm{ext}} - \mu_{\mathrm{A}}^*}{RT}\right)\right] \tag{6.18}$$

where $\alpha = a\nu \exp\left(-\dfrac{Q_b}{RT}\right)$ is the interfacial reaction parameter. If the chemical potential is due to entropy only, then $\mu_{\mathrm{ext}} = RT \ln(c_{\mathrm{ext}})$ and $\mu_{\mathrm{A}}^* = RT \ln(c_{\mathrm{A}}^*)$ and equation (6.18) simplifies to

$$j^* = \alpha[c_{\mathrm{ext}} - c_{\mathrm{A}}^*]. \tag{6.19}$$

If $c_{\mathrm{A}}^* < c_{\mathrm{ext}}$ then material flows across the interface from the reservoir into the specimen. If $c_{\mathrm{A}}^* > c_{\mathrm{ext}}$ then material flows from the specimen into the reservoir. This boundary condition therefore always acts to drive the interfacial concentration in the specimen c_{A}^* towards the concentration of the species in the reservoir, c_{ext}.

The effect of boundary condition choice is illustrated in figure 6.6. For the following examples, the linearised diffusion equation of equation (6.15) is solved in one-dimension

$$\frac{\partial c_{\mathrm{A}}}{\partial t} = D_{\mathrm{A}}\frac{\partial^2 c_{\mathrm{A}}}{\partial x^2} \quad \text{for} \quad 0 \leqslant x \leqslant L$$

where $L = 1$ mm. The tracer diffusion constant is taken to be $D_{\mathrm{A}} = 10^{-11}$ m^2 s^{-1}. The initial condition is taken to be $c_{\mathrm{A}}(x, 0) = 0.0046$. The left-hand boundary condition is varied and a zero flux condition is applied to the right-hand boundary, i.e. $j(L,t) = 0$. The diffusion problem can be solved using many numerical packages, including Excel for simple 1D problems such as this. Here it is solved using the finite element software COMSOL Multiphysics v5.5 [7]. Figure 6.6(a) demonstrates the Dirichlet condition. At time $t = 0$ there is an incompatibility between the initial condition, $c_{\mathrm{A}}(0, 0) = 0.0046$, and the boundary condition, $c_{\mathrm{A}}(0, t) = 0.075$, leading to a step change in concentration. This high (infinite) concentration gradient rapidly disappears and diffusion slows at the concentration gradually diminishes to zero, at which point the final concentration field is uniform at the prescribed boundary value. Figure 6.6(b) illustrates the Neumann condition. In this case the system never settles as material continues to enter at a fixed rate and hence must continue to be transported away from the boundary. There is no upper limit on the concentration in this case, i.e. the unphysical condition of $c_{\mathrm{A}} > 1$ can be reached. Figures 6.6(c) and (d) illustrate the Robin condition. In the first example, the interface reaction parameter (α) is large enough such that the smaller diffusion parameter (D_{A}) controls the process, i.e. additional material from the external reservoir easily enters the system. The second example has a lower interfacial reaction parameter. Now diffusion is the faster process, demonstrated by the flat concentration profiles, and the interfacial reaction is controlling the rate at which the process evolves. A more practical (non-linear) diffusion problem is addressed in the next sub-section.

6.5.1 An example non-linear diffusion problem: carburisation of a ferritic steel

Carburisation is the ingress of carbon into a material from an external source. It is often used to raise the carbon content near the surface of low carbon steels to increase their surface hardness. It can also be detrimental to stainless steels in a high temperature and pressure CO_2 environment, where excessive carburisation over long

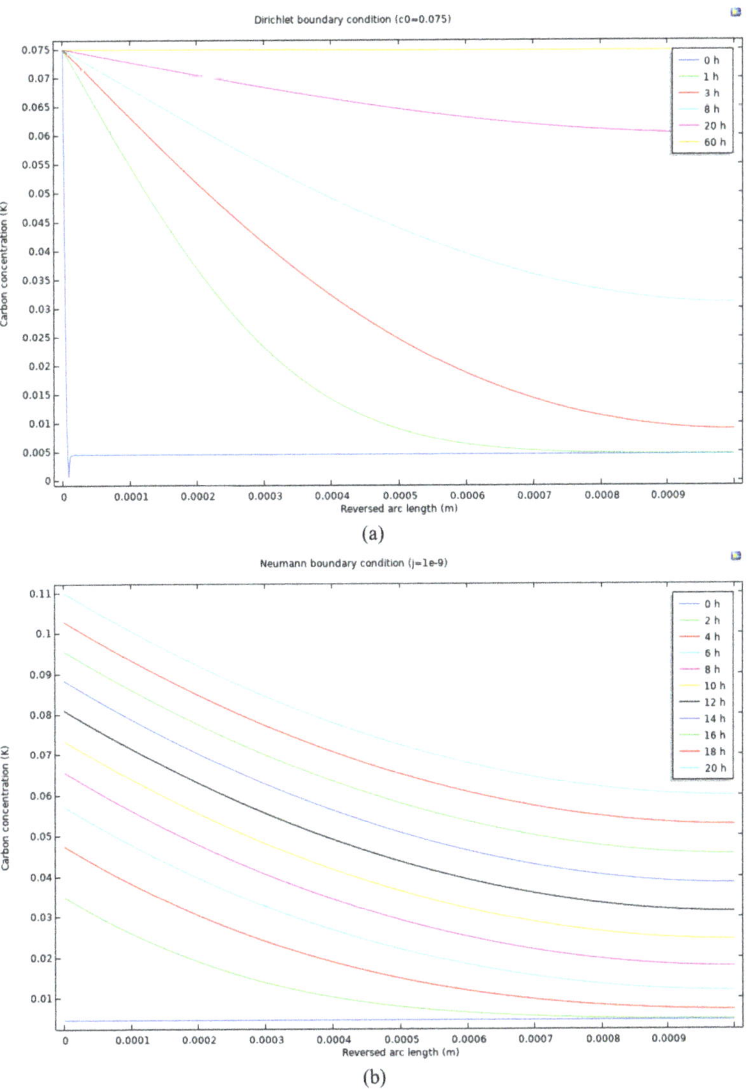

Figure 6.6. The concentration profile $c_A(x, t)$ through the sample as a function of time for different boundary conditions. The horizontal axis is in metres (from 0 to 1 mm). Note the different time scales between images. (a) Dirichlet boundary condition $c_0^* = 0.075$, (b) Neumann boundary condition $j_0^* = 10^{-9}$ m s^{-1}, (c) Robin boundary condition with $c_{ext} = 0.075$ and $\alpha = 10^{-7}$ m s^{-1}, and (d) Robin boundary condition with $c_{ext} = 0.075$ and $\alpha = 10^{-9}$ m s^{-1}. Modelling performed using COMSOL Multiphysics® [7].

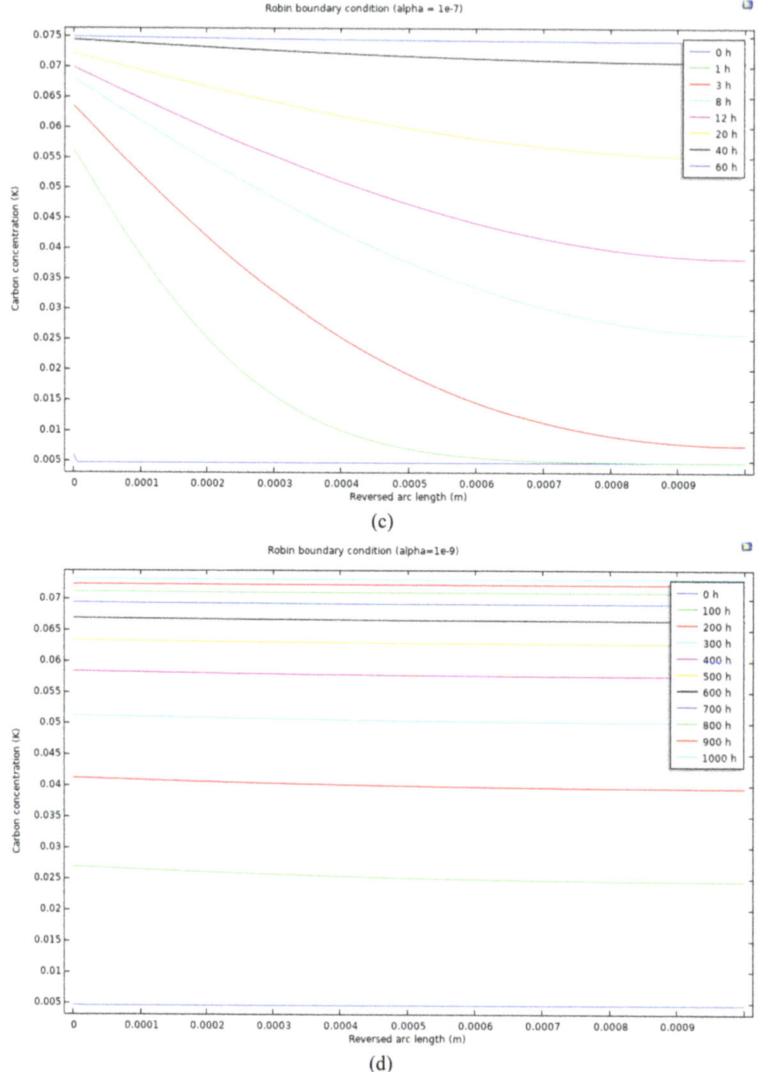

(c)

(d)

Figure 6.6. (Continued.)

time periods can lead to embrittlement. Here we consider the latter case where the carburisation process is slow. Here the substitutional carbide formers, mainly Cr, have time to locally diffuse, allowing Cr-based carbides to form. As such it is reasonable to assume that the material reaches local equilibrium at every point. The steel is Fe–9Cr–1Mo which is 9 wt% Cr, 1 wt% Mo and 0.1 wt% C (or 9.6 at% Cr, 0.58 at% Mo and 0.46 at% C) with a ferrite (BCC phase) matrix. Carbon is the only interstitial atom in the matrix and is expected to diffuse very quickly relative to the substitutional elements (Fe, Cr and Mo). Consequently only the long range diffusion of carbon is considered, and the substitutional elements are assumed to achieve the short range diffusion required to form carbides within the time available. The local carbon content, $c_C = c_m + c_p$, is divided between the carbon in the matrix, c_m, and the carbon in the carbide precipitates, c_p. Ferrite is well know for having a very low solubility for carbon, such that $c_m \ll c_p$. Hence we can assume that $c_C \approx c_p$.

We solve equation (6.14) in one-dimension

$$\frac{\partial c_C}{\partial t} = \frac{\partial}{\partial x}\left(D_C^{BCC*}\frac{\partial c_C}{\partial x}\right) \text{ for } 0 \leqslant x \leqslant L \tag{6.20}$$

with $L = 1$ mm, subject to the initial condition of $c_C(x, 0) = 0.0046$, and the boundary conditions

$$j(0, t) = \alpha(c_{ext} - c_C^*) \text{ and } j(L, t) = 0 \tag{6.21}$$

where the former uses the linearized version of the Robin boundary condition, equation (6.19), to represent the exposure of the free surface to an influx of carbon from the CO_2 atmosphere, with $c_{ext} = 0.075$ and $\alpha = 10^{-10}$ m s^{-1}. The diffusivity

$$D_C^{BCC*} = \frac{c_m D_C^{BCC}}{RT}\frac{\partial \mu_C}{\partial c_C} \tag{6.22}$$

is highly non-linear. Diffusion is through the ferrite matrix so c_m represents the probability of a carbon atom being available for a diffusive hop. As local equilibrium is assumed at every point, the chemical potential of carbon, $\mu_C(c_C, T)$, is the same in both the matrix and the carbide phases (see section 3.2.2). The tracer diffusivity of carbon in ferrite [6] is given by

$$D_C^{BCC}(T) = 6.2 \times 10^{-7}\exp\left(-\frac{80 \times 10^3}{RT}\right) \text{[m}^2 \text{ /s}^{-1}\text{]}.$$

A uniform temperature of 600 °C is assumed, giving $D_C^{BCC}(600\,°C) = 10^{-11}$ m^2 /s^{-1}. The equilibrium quantities μ_C and c_m can be calculated as a function of carbon concentration and temperature using any thermodynamic package with the appropriate TDB file, such as PANDAT [8], Thermocalc [9] or JMatPro [10]. The thermodynamics for this four phase system were determined using JMatPro [10] and are shown in figure 6.7. The mole fraction of equilibrium phases is shown in figure 6.7(a) as a function of the total carbon content at a point, with the ferrite matrix making up the remaining fraction. As the carbon content of the matrix is very low, the bulk of the carbon forms carbides. Initially these are $M_{23}C_6$, where the M denotes the metallic elements. In practice these are largely $Cr_{23}C_6$ carbides. M_7C_3 carbides start to form above 2.6 at% C and are the majority precipitate phase above 4 at% C. The cementite phase (Fe_3C) appears above 8.2 at% C, although as this is above the source concentration $c_{ext} = 7.5$ at% so this is not relevant to the calculation. The chemical potential, $\mu_C(c_C, 600\,°C)$, is plotted in figure 6.7(b), showing a steady rise with carbon content until it levels off with the appearance of cementite at 8.2 at%. This flattening of the curve is also seen in the matrix carbon

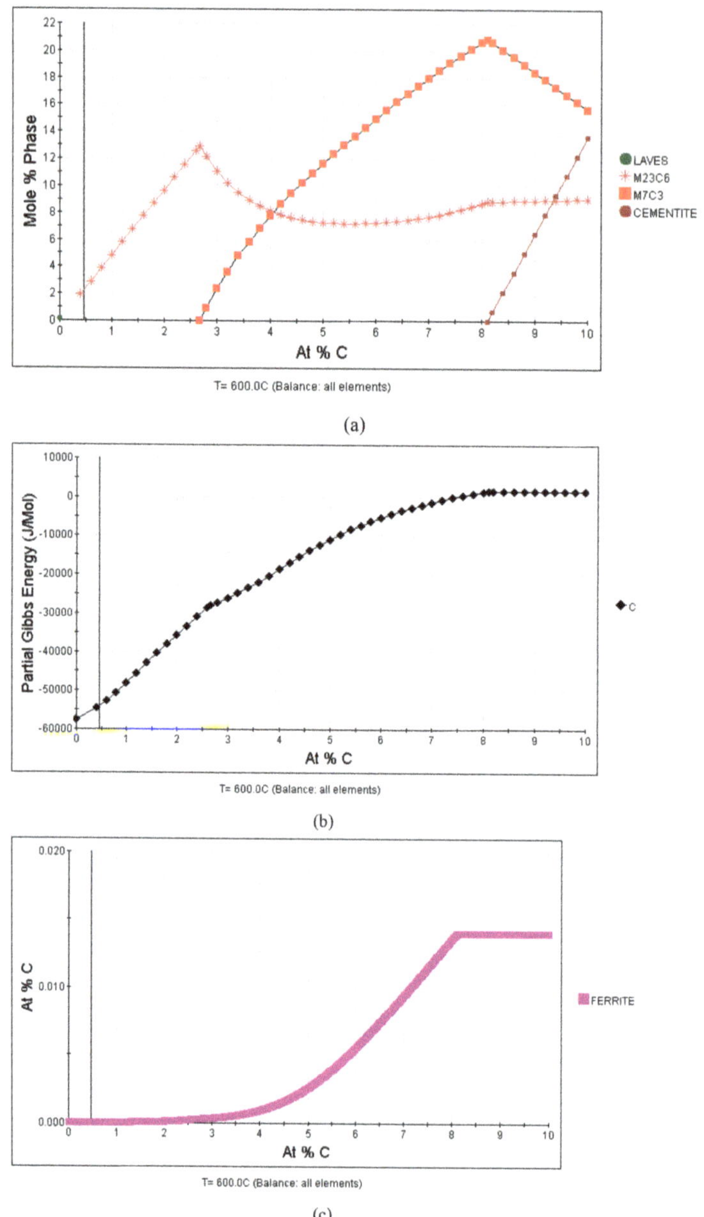

Figure 6.7. Thermodynamic properties for Fe–9Cr–1Mo–C at 600 °C, created using the software JMatPro® [10]: (a) the equilibrium phases as a function of carbon concentration (atomic %), (b) the chemical potential $\mu_C(c_C)$ and (c) the carbon concentration in the ferrite matrix, $c_m(c_C)$.

concentration, $c_M(c_C, 600\,°C)$, shown in figure 6.7(c), where all additional carbon is entering the cementite phase.

The diffusion problem defined by equations (6.20) and (6.21) is solved using COMSOL Multiphysics v5.5 [7]. The data from figures 6.7(b) and (c) was used to calculate the chemical diffusion coefficient $D_C^{BCC*}(c_C, 600\,°C)$ of equation (6.22). This data was then loaded into a lookup table to define a linear interpolation function for the diffusivity as a function of carbon concentration. The resulting function is illustrated in figure 6.8(a). There are a number of interesting points to note. Firstly, the diffusivity is a very strong function of the carbon concentration,

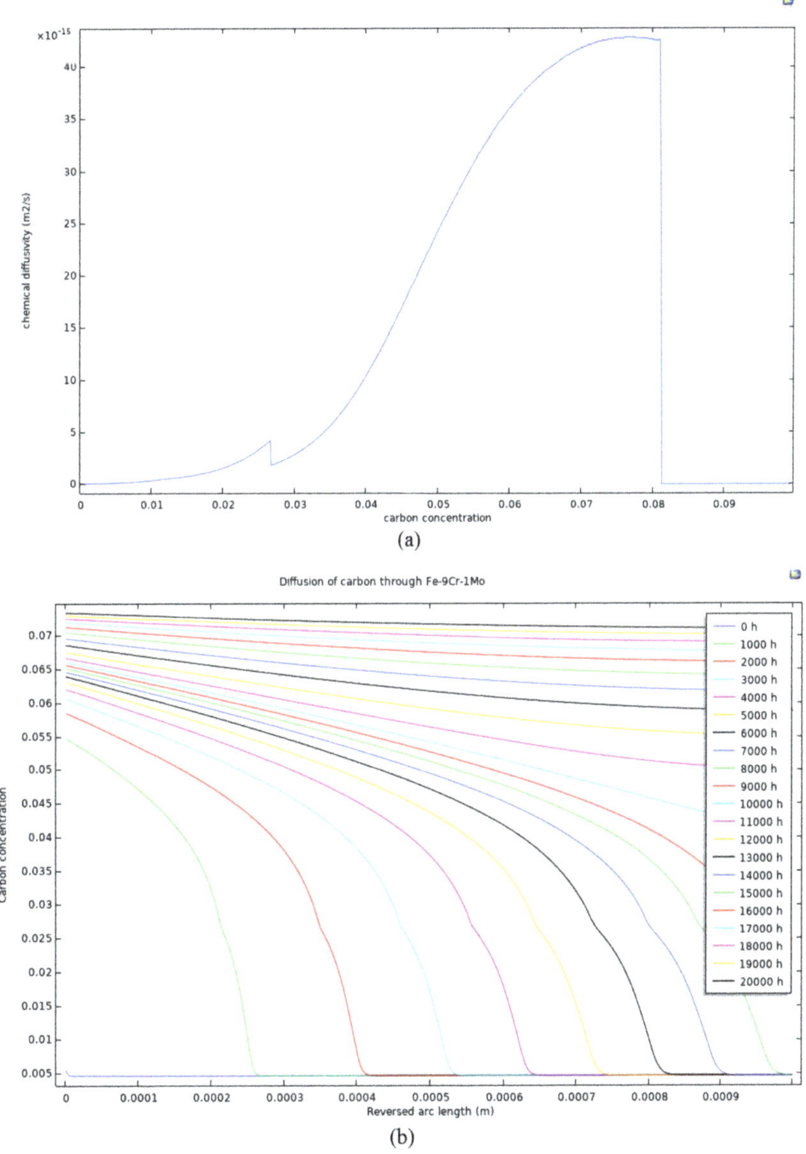

(a)

(b)

Figure 6.8. (a) The chemical diffusivity $D_C^{BCC*}(c_C, 600\,°C)$ as a function of the carbon concentration c_C (mole fraction). (b) The carbon concentration profile $c_C(x, t)$ through the sample as a function of time. The horizontal axis is in metres (from 0 to 1 mm). Modelling performed using COMSOL Multiphysics® [7].

changing by three orders of magnitude over the given range. The diffusivity increases rapidly with carbon concentration, but with two sharp discontinuities, associated with a sudden change in the chemical potential gradient. The first is associated with the appearance of the M_7C_3 carbides and the second with the appearance of cementite, for which the flat chemical potential curve results in zero diffusivity. Overall the chemical diffusivity ($10^{-17} - 10^{-14}$ m^2 s^{-1}) is very much lower than the tracer diffusivity (10^{-11} m^2 s^{-1}) used for the simplified purely Fickian models of figure 6.6.

Concentration profiles for the carbon content through the steel sample are shown in figure 6.8(b) at intervals of 1000h. The non-linearity of the diffusion coefficient can be seen in the early carbon profiles (where the carbon content is low) as their curvature is in the opposite direction to those seen in linear diffusion models (see figure 6.6). There is also a kink in the profiles at a carbon concentration of 0.026 (2.6at%) due to the discontinuity in the diffusion coefficient at this value (see figure 6.8(a)). Diffusion is slower at lower carbon levels, and this largely controls the initial carbon ingress, with the carbon penetrating the sample with a well-defined interface until the right-hand boundary is met at around 8000h. At this point the carbon level is fairly high everywhere and diffusion can occur more rapidly. This causes the carbon profiles to flatten and the interfacial reaction rate (determined by the kinetic constant, α) becomes the rate limiting process.

References

[1] Carslaw H S and Jaeger J C 1959 *Conduction of Heat in Solids* 2nd edn (London: Oxford Press)

[2] Shewman P G 1963 *Diffusion in Solids* (New York: McGraw-Hill)

[3] Majumdar P 2005 *Computational Methods for Heat and Mass Transfer* (Boca Raton, FL: CRC Press)

[4] Logan D 2016 *A First Course in the Finite Element Method* (CL Engineering)

[5] Reddy J N and Gartling D K 2010 *The Finite Element Method in Heat Transfer and Fluid Dynamics* (Boca Raton, FL: CRC Press)

[6] Bhadeshia H and Honeycombe R 2017 *Microstructure and Properties* 4th edn (Oxford: Butterworth-Heinemann)

[7] COMSOL Multiphysics® v. 5.6. COMSOL AB, Stockholm, Sweden www.comsol.com [accessed 1 October 2021]

[8] Pandat: software suite for thermodynamic calculation and kinetic simulation of multi-component alloys. CompuTherm LLC, Madison, Wisconsin, USA www.computherm.com [accessed 1 October 2021]

[9] Thermocalc software: https://thermocalc.com/ [accessed 1 October 2021]

[10] JMatPro®, a software developed by Sente Software Ltd (UK) www.sentesoftware.co.uk/jmatpro [accessed 1 October 2021]

IOP Publishing

Thermodynamics, Kinetics and Microstructure Modelling

Simon P A Gill

Chapter 7

Nucleation, growth and coarsening in solids

The appearance of a new stable phase is characterised by nucleation followed by growth. This applies to nucleation of a solid phase in a liquid, or a solid phase in a solid. However, here we focus on the growth of a solid precipitate phase in a solid matrix. The growth of a solid phase in a liquid, known as solidification, is different in a number of ways and is covered independently in chapter 9. In a solid, if nucleation is fast, and growth is slow, then a fine dispersion of small precipitate phases results. Conversely, if nucleation is slow and growth is fast, then the precipitate phase consists of a few large particles a large distance apart. As discussed in section 1.3, these microstructural differences have significant consequences for the final properties of the material. Simple models for the processes of nucleation and stable growth are introduced here to aid in the prediction and understanding of how microstructures are created. Finally, the interfacial energy between solid phases can be further minimised after growth has completed by coarsening of the precipitate dispersion. A model for this process is introduced in section 7.4.

7.1 Nucleation

The energetics of nucleation, introduced in section 5.3, prescribe a critical radius, r^*, given by equation (5.11), for the appearance of stable nuclei. Atomic clusters above this size will grow (are stable) and clusters below this size will shrink (are unstable). There is an energy increase associated with the appearance of a nucleus, ΔG^*, which is a competition between the reduction in energy due to the phase transformation from the host phase α to the nucleus phase β, $\Delta g^{\text{nuc}}_{\alpha \to \beta}$, given by equation (5.4), and the rise in energy due to the creation of the α/β interface, characterised by the interfacial energy per unit area, $\gamma_{\alpha\beta}$. The equilibrium number of critical clusters that form in a phase containing N_0 atoms is given by

$$N_{\text{nuc}} = N_0 \exp\left(-\frac{\Delta G^*}{k_B T}\right)$$

based on the probability of this configuration occurring due its higher energy state, where $k_B = 1.381 \times 10^{-23}$ J K^{-1} is Boltzmann's constant. Nucleation, however, also requires diffusional transport of material to form a nuclei at a point, and hence the actual number of nuclei also depends on kinetic factors. It is more useful, therefore, to consider the nucleation rate

$$\dot{N}_{\text{nuc}} = f_0 N_0 \exp\left(-\frac{\Delta G^*}{k_B T}\right) \tag{7.1}$$

where $f_0 = Z\beta^*$ is the frequency at which atoms attempt to form a nuclei. This has two components: the Zeldovich factor, Z, is the probability that a nucleus that has sufficient activation energy will go on to be a stable size and not dissolve; and β^*, which is the rate at which atoms attach to the nucleus causing it to grow. This latter quantity depends on D^α, the tracer diffusivity of the relevant component in the host phase. As the units of D^α are m^2 s^{-1} we can write $\beta^* = D^\alpha/\lambda^2$ where λ is a characteristic length for the diffusion process. There are a number of theories for estimating f_0 and its components, but they are generally incorrect by a number of orders of magnitude. They are not elaborated further here for this reason, and in practice f_0 is typically determined experimentally. This demonstrates how complex the nucleation process can be and how many factors are involved.

However, it is useful to investigate the predictions of equation (7.1) in relation to the fictitious AB system of section 5.2, to gain an appreciation of how the different physical processes involved define the overall nucleation response. For continuity we consider the nucleation of the solid (S) phase in the liquid (L) phase for a composition of A of $c_B = 0.2$ for which the melting temperature is $T^M = 722\,°C$, as shown in figure 5.2(a). The associated energetics are used to calculate the thermodynamic component, $\exp\left(-\dfrac{\Delta G^*}{k_B T}\right)$, of the nucleation rate, shown in figure 7.1(a). Both the homogeneous case, given by equation (5.13) and shown in figure 5.4, and the heterogeneous case for precipitation on a surface (in a liquid) or on a grain boundary (in a solid) with a dilation angle of $\psi = 60°$, given by equation (5.19), are shown. This is the probability for a nuclei to appear, and it can be seen that it is zero at the melting temperature, but increases substantially when the system is undercooled. The energy barrier to heterogeneous nucleation is much lower than that for homogeneous nucleation and hence this is more probable. The kinetic component of the nucleation rate f_0 is estimated in figure 7.1(b) based on a diffusivity with a typical activation energy $Q_A^\alpha = 240$ kJ mol^{-1}. The diffusion distance is taken to be proportional to a precipitates size, which introduces a small difference between the homogeneous and heterogeneous cases, but is not particularly significant. It can be seen that the kinetic component increases with temperature as one would expect.

The overall nucleation rate, shown in figure 7.1(c), is the product of the two terms shown in figures 7.1(a) and (b) and shows that nucleation is a competition between kinetics and thermodynamics. The thermodynamics favours large undercooling, whereas the kinetics slows as the temperature drops. This competition leads to a

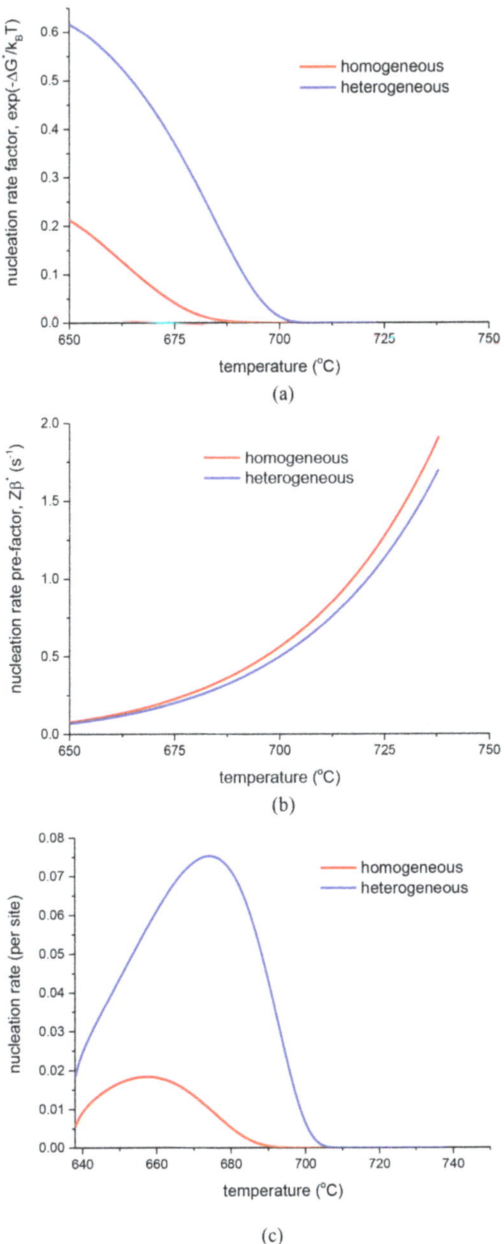

Figure 7.1. The nucleation rate of equation (7.1) for the **AB** system of section 5.2: (a) thermodynamic component, (b) kinetic component, (c) the net nucleation rate is the product of the thermodynamics and kinetic factors.

characteristic peak in the nucleation rate at a critical undercooling. This shows that different levels of undercooling can be used to influence the resulting microstructure. Heterogeneous nucleation is always much easier than homogeneous nucleation, and homogeneous nucleation is rare for that reason. During the solidification of a liquid, the nuclei will always form on the walls of the container or on the surface of impurity particles, sometimes known as seed particles if they are introduced deliberately to facilitate nucleation. Intragranular precipitates will generally nucleate on a defect in a crystal, such as a dislocation or grain boundary.

7.2 Growth of a solid phase in a solid matrix

Once a stable nucleus has formed, it will start to grow. This generally requires transport of material, either towards it or away from it. This diffusional transport controls the rate of growth, and this is therefore known as **diffusion-controlled growth**. This is considered in section 7.2.2, but first the less common case of diffusionless **interface-controlled growth** is introduced as a first step towards developing this model.

7.2.1 Interface-controlled growth

In a few cases, a phase transformation occurs without need for redistribution of the solute. This is a diffusionless transformation. One such case is the transformation in steel of austenite into martensite. The martensitic transformation is known as a **displacive transformation**, as the cubic unit cell of austenite undergoes a rapid distortion into the asymmetric unit cell of martensite, which is slightly longer in one direction and shorter in the other two. In this case, the progress of the transformation is determined by the mobility of the interface, M_{in}. The velocity of the interface is then proportional to the driving force acting on it, i.e. $v = M_{in}f$, where f is the driving force per unit area of interface. The following model derivation introduces the variational approach of equation (1.1) to solving diffusional problems [1].

Derivation of interface-controlled growth rate

To determine the growth rate, we compare the rate of change of energy with the rate of energy dissipation. The rate of change of energy provides the driving force for growth for a spherical phase of radius r

$$\Delta \dot{G} = \frac{\partial \Delta G}{\partial r}\dot{r} = -F\dot{r} \tag{7.2}$$

where the driving force F is given by equation (5.12). This is the power consumed in doing work, i.e. power is force times velocity. Power is dissipated by movement of the interface and is again a product of force times velocity such that the dissipation potential is given by

$$\Psi = \frac{1}{2}\int_A vf\,\mathrm{d}A = \frac{1}{2}\int_A \frac{v^2}{M_{in}}\mathrm{d}A = \frac{1}{2}\cdot 4\pi r^2 \cdot \frac{\dot{r}^2}{M_{in}} \tag{7.3}$$

where $A = 4\pi r^2$ is the interfacial area. The total power according to equation (1.1) is

$$\Pi = \psi + \Delta\dot{G}. \tag{7.4}$$

Substituting equations (7.2) and (7.3), the velocity adopted by the interface is that which optimises the power dissipated with respect to the power released according to equation (1.2) such that

$$\frac{d\Pi}{d\dot{r}} = 4\pi r^2 . \frac{\dot{r}}{M_{\text{in}}} - F = 0$$

giving

$$\dot{r} = M_{\text{in}} f \tag{7.5}$$

where $f = \dfrac{F}{4\pi r^2}$ is the driving force per unit area.

The resulting growth rate of the particle from equations (7.4) with (5.12) is

$$\dot{r} = 2\gamma M_{\text{in}}\left[\frac{1}{r^*} - \frac{1}{r}\right]. \tag{7.6}$$

It can be seen that growth only occurs for nuclei with $r > r^*$ as expected. The growth is only slowed by the increase in interfacial energy, but this contribution reduces as the particle gets large. Most displacive transformations are known as massive transformations, whereby a significant percentage (>80%) of the system is transformed. In this case the growing spherical phases will start to overlap. This sort of massive transformation can be taken into account using the Avrami equation, which will be introduced in section 8.3.

7.2.2 Diffusion-controlled growth

Most phase transformations are diffusion controlled. In this section, we focus on the precipitation of a minor solid phase within a solid matrix. A minor transformation is defined as one where the precipitation volume fraction is sufficiently small (<20% say) such that the precipitates do not impinge upon one another. Solidification from a liquid into a solid has many similarities to this, but is generally much faster, leading to complex growth morphologies. This case is therefore considered separately in chapter 9.

In general there is a difference between the equilibrium solute composition in the spherical precipitate phase, c_β, and that in the host matrix phase, c_α. This is illustrated in figure 7.2, where the radial compositional profile across the precipitate/matrix interface, shown as a blue line, is approximated by a linear variation for the purposes of this model. The lever rule of equation (3.19) tells us that one of these must be greater than the nominal composition, c_0, and one must be less. It is common to assume that the less abundant component in the solid is the reference component, i.e. we are on the left side of the phase diagram (see figure 5.2(a), in which the SOLID is the precipitate phase, the LIQUID is the host phase and the solute is component B). Hence we have $c_\beta < c_0$ such that solute atoms must be ejected from the precipitate phase as it grows. Chemical equilibrium is achieved at the interface, such that the matrix composition there is c_α. We necessarily have $c_\alpha > c_0$ so there is a negative compositional gradient taking solute away from the

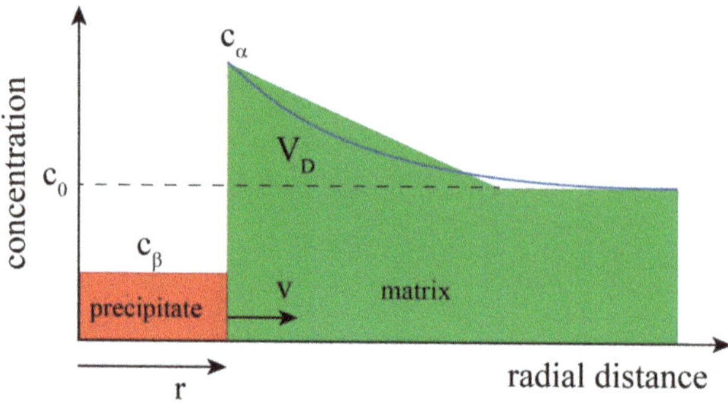

Figure 7.2. Schematic of diffusion-controlled growth. As the precipitate (red) grows it expels solute material into the surrounding matrix (green) leading to a pile-up of solute ahead of the particle which slows its growth as its size increases. The real concentration profile is smooth, as indicated by the blue line, but here it is approximated by a straight line for the derivation of the variational model in equation (7.10).

interface so that the phase transformation can proceed. Recall that material diffuses down a concentration gradient, according to equation (6.9). The rate at which the solute is removed from the interface strongly influences the possible growth rate of the precipitate. Initially the concentration gradient will be large and growth can occur relatively quickly. However, as solute builds up in front of the interface, the gradient will lessen, and the growth will slow. An approximate expression for the growth rate is determined in the following derivation.

Derivation of diffusion-controlled growth rate

An advantage of the variational approach is that it does not solve the diffusion equations exactly at every point, but instead balances the total dissipation rate against the total driving force, and hence can be used to develop useful approximations to problems with simplified concentration profiles. Here we assume the concentration profile is linear, as shown in figure 7.2. The rate of movement of solute atoms is defined by the volumetric flux, j. This is the volume of solute crossing a cross-section of unit area per unit time, as introduced in section 6.3. The material flux $j = D_f F_m$ is proportional to the driving force (per mole) for movement of the solute, as given in equation (6.3), where $D_f = \dfrac{c_0 D_B^\alpha}{RT}$ depends on the diffusion constant for solute B in the matrix phase α. The power dissipated by this process is equal to the rate of movement (flux) times the driving force (per unit volume), $F_V = F_m/\Omega$, such that

$$\Psi = \frac{1}{2} \int_V F_V j\, dV = \frac{1}{2} \int_V \frac{j^2}{\Omega D_f}\, dV \qquad (7.7)$$

where V is the volume of the α phase. The rate of solute removal from (or addition to) the precipitate must be equal to the flux. For a discrete jump in the concentration at an interface one can write equation (6.12) in discrete form as

$$\frac{\Delta c}{\Delta t} = -\frac{\Delta j}{\Delta x}.$$

This can be rearranged to

$$\Delta j = \Delta c v_n$$

where $v_n = \Delta x / \Delta t$ is the normal velocity of the interface, and $\Delta c = c_\beta - c_\alpha$ is the difference in concentration across the interface. As there is no flux in the precipitate we can write $\Delta j = j = \Delta c \dot{r}$, where $v_n = \dot{r}$. Thus the dissipation potential of equation (7.7) is

$$\Psi = \frac{(\Delta c^2 \dot{r}^2)}{2 D_f \Omega} V_D \tag{7.8}$$

where V_D, the volume within which diffusion occurs, is to be determined. Mass conservation requires that the volume of material expelled from the precipitate must equal the volume added to the diffusion zone. For planar growth this gives

$$(c_0 - c_\beta) V_\beta = \frac{1}{2}(c_\alpha - c_0) V_D. \tag{7.9}$$

For spherical growth the solution is more complicated, but we retain this approximation here for simplicity. Here $V_\beta = \frac{4}{3}\pi r^3$ is the volume of the precipitate phase. Substituting equation (7.9) into (7.8) gives

$$\Psi = \frac{(\Delta c^2 \dot{r}^2)}{2 D_f \Omega} \cdot \frac{2(c_0 - c_\beta)}{(c_\alpha - c_0)} \cdot \frac{4}{3}\pi r^3.$$

The power consumed follows equation (7.4) with the driving force of equation (7.2) as before. The optimal growth rate minimises the functional such that

$$\frac{d\Pi}{d\dot{r}} = 4\pi r^2 \cdot \frac{2(c_0 - c_\beta)}{3(c_\alpha - c_0)} \cdot \Delta c^2 r \frac{\dot{r}}{D_B^\alpha \Omega} - F = 0.$$

Hence

$$\dot{r} = \frac{b D_f \Omega}{\Delta c^2} \cdot \frac{f}{r}$$

where the constant

$$b = \frac{3(c_\alpha - c_0)}{2(c_0 - c_\beta)}$$

is an approximation for spherical growth. In general, these idealised models are very useful to conceptualise the important factors governing growth, and the relevant scaling relationships, but they cannot be expected to be highly accurate, especially when many of the material input properties are not well characterised. Hence we acknowledge the level of accuracy of the model by assuming that $b = 1$, which is similar to the approach taken in the PanPrecipitation model in PANDAT.

The growth rate of a particle undergoing diffusion-controlled growth is therefore given by

$$\dot{r} = \frac{K}{r}\left[\frac{1}{r^*} - \frac{1}{r}\right] \tag{7.10}$$

where the rate constant is given by $K = \dfrac{2\gamma D_{\mathrm{B}}^{\alpha}\Omega c_0}{\Delta c^2 RT}$ using equation (5.12) and taking $b = 1$. The main difference between the interface-controlled growth rate of equation (7.6) and the diffusion-controlled growth rate of equation (7.10) is that the latter has an additional $1/r$ factor. This slows the diffusion-controlled growth rate as the particle gets larger. This is because, as mentioned before, the solute ejected from the precipitate piles up in front of the moving interface as the particle grows, slowing its growth rate as it gets larger.

7.3 Coarsening

Growth of particles will eventually stop once the equilibrium volume fraction of the precipitating phase is reached. However, the evolution of the system does not necessarily stop at that point, if the material remains at a sufficiently elevated temperature such that diffusion can occur. This is because the interfacial energy can still be reduced by further changes in the size and distribution of precipitates. It is easy to see this if you calculate the interfacial area of one large sphere of radius a compared to the interfacial areas of N spheres of radius c with the same total volume such that

$$V = \frac{4}{3}\pi a^3 = N\frac{4}{3}\pi c^3$$

giving $\dfrac{a}{c} = N^{\frac{1}{3}}$. The ratio of the surface (interfacial) area of the large sphere to the N small spheres is

$$\eta = \frac{4\pi a^2}{N4\pi c^2} = \frac{1}{N}\left(\frac{a}{c}\right)^2 = N^{-\frac{1}{3}}$$

showing that the surface area of the smaller spheres is always bigger than the single large sphere and increases with N. It is therefore clear that there is a driving force to reduce the total interfacial energy (area) by increasing the mean size of the particles. This process is known as **coarsening**, or **Ostwald ripening** [2], and is usually detrimental to a materials mechanical performance, which prefers a fine dispersion of precipitates (see section 1.3). As the total volume of particles is fixed by thermodynamic equilibrium, coarsening can only be achieved by some (larger) particles consuming neighbouring (smaller) particles, as shown in figure 7.3.

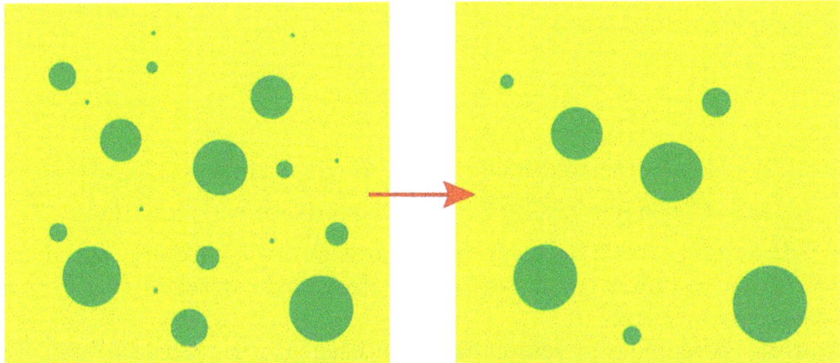

Figure 7.3. Reduction in interfacial energy drives the coarsening of a microstructure. A finer dispersion of precipitates (left) will gradually coarsen over time at elevated temperature, causing the number of particles to decrease and the average size of the remaining particles to increase. This occurs by transfer of material between smaller particles and larger particles.

7.3.1 A globally averaged model for coarsening

The general scaling law for the evolution of the mean particle size due coarsening is derived as follows.

Derivation of coarsening rate

The aim is to calculate how the mean particle volume, $\overline{V}$, changes over time. Let the total constant precipitate volume be V_{TOT}. Then $N(t)$, the number of particles at time t, satisfies

$$\overline{V} N = V_{\text{TOT}}.$$

Taking the time derivative, and given $\dot{V}_{\text{TOT}} = 0$, we have

$$\dot{\overline{V}} = -\frac{\dot{N}}{N}\overline{V} \tag{7.11}$$

where it can be seen that the mean volume only increases if the number of particles decreases, i.e. $\dot{N} < 0$. Now consider the behaviour of an individual particle, with radius r_i, such that its volume is $V_i = \frac{4}{3}\pi r_i^3$. The growth rate of equation (7.10) also applies for the post-growth, coarsening rate of this particle and gives

$$\dot{V}_i = 4\pi r_i^2 \dot{r}_i = 4\pi K\left[\frac{r_i}{r^*} - 1\right]. \tag{7.12}$$

We know that the rate of change of the total volume of the precipitate phase must be zero so that, summing over all particles we get

$$\dot{V}_{\text{TOT}} = \sum_N^{i=1} \dot{V}_i = 4\pi K \sum_N^{i=1}\left[\frac{r_i}{r^*} - 1\right] = 0.$$

This condition is satisfied when

$$r^* = \frac{1}{N}\sum_i r_i = \bar{r}$$

i.e. the critical radius in the coarsening regime is equal to the mean particle radius, $\bar{r}$, where $\bar{V} = \frac{4}{3}\pi\bar{r}^3$. The simple result of this, in accordance with equation (7.12), is that particles with a radius above the mean radius grow during coarsening, and those with a size below the mean radius shrink, and eventually disappear, leading to a net increase in the average particle size.

The rate of coarsening is therefore the rate at which the (smaller) particles, for which $r_i < \bar{r}$, disappear. We can determine the time a particle of radius $r_i = s_i\bar{r}$ takes to disappear by solving equation (7.12), where we require the constant $0 \leqslant s_i < 1$ for a shrinking particle, such that

$$\int_0^{s_i\bar{r}} \frac{\bar{r}r_i^2}{r_i - \bar{r}}\mathrm{d}r_i = A(s_i)\bar{r}^3 = K\tau_i \tag{7.13}$$

where $A(s) = -\left[s + \frac{1}{2}s^2 + \ln(1 - s)\right]$ is a positive size-dependent constant and τ_i is the time it takes for particle i to disappear. The limiting value $A(1) = \infty$ shows that a particle with the mean radius does not disappear, i.e. it shrinks infinitesimally slowly. This also demonstrates that the time for any small particle to disappear scales as $\bar{r}^3$. The disappearance rate for the shrinking particles is obtained by summing over the shrinkage rates for particles with $s_i < 1$

$$\dot{N} = -\sum_{s_i<1}\frac{1}{\tau_i} = -\frac{K}{\bar{r}^3}\sum_{s_i<1}\frac{1}{A(s_i)}.$$

Combined with equation (7.11) this yields

$$\dot{\bar{V}} = \frac{4}{3}\pi Kd \tag{7.14}$$

where $d = \frac{1}{N}\sum_{s_i<1}\frac{1}{A(s_i)}$. It is assumed that the particle size distribution reaches a self-similar state, whereby it does not change shape over time, but just rescales relative to the evolution of the mean radius. In this case d is a dimensionless constant, and hence equation (7.14) predicts that the mean particle volume increases at a constant rate over time.

The overall rate of coarsening, given by equation (7.14) is more commonly expressed as

$$\bar{r}^3(t) - \bar{r}^3(0) = k_c t \tag{7.15}$$

where the coarsening constant $k_c = Kd$. In the original derivation, Lifshitz and Slyozov [2] determined the steady state particle size distribution exactly, which yielded

$k_\mathrm{c} = \dfrac{8\gamma D_\mathrm{B}^\alpha \Omega c_0}{9RT} = \dfrac{4K\Delta c^2}{9}$, suggesting $d = \dfrac{4\Delta c^2}{9}$. Given (7.15), the evolution of a particle dispersion can be predicted as a function of time and temperature and related to changes in material performance through relationships such as those in section 1.3.

7.3.2 A local mechanism for coarsening

So far we have considered the global driving force for coarsening (reduction in interfacial energy) and deduced that the mean particle size evolves on average over time according to equation (7.15). But how does a particle know if it is expected to shrink (below the average particle size) or grow (above the average particle size)? Particles coarsen by diffusion of solute between each other. If a particle ejects solute as it grows, then a shrinking particle will need to absorb solute back into itself to do so. Hence, in this case, there must be diffusion of solute from the big (growing) particles into the small (shrinking) particles. And, as such, there must be a concentration gradient between particles for the solute to diffuse down, with a higher solute concentration around big particles and a lower solute concentration around smaller particles. This dependence of the interfacial solute concentration on particle size is known as the **Gibbs–Thompson effect**. The origins of this interfacial effect have already been captured in the particle growth rate of equation (7.10), which came about from the addition of the interfacial energy term to the Gibbs free energy of equation (5.9). Here the effects of these energetics on the local interactions between particles is made explicit.

Derivation of Gibbs–Thompson effect

The equilibrium conditions for a two phase mixture in the absence of interfacial energy were derived in section 3.2.2, and are given for a solid (S) and liquid (L) mixture in equation (3.17). Here equation (3.17) is adopted for the solid–solid transition from the α matrix phase to the β particle phase such that

$$\Delta\mu^{\alpha*} = \Delta\mu^{\beta*} = \frac{g_\alpha\!\left(c_B^{\alpha*}\right) - g_\beta\!\left(c_B^{\beta*}\right)}{c_B^{\alpha*} - c_B^{\beta*}} = \frac{\Delta g_{\alpha\to\beta}}{c_B^{\alpha*} - c_B^{\beta*}} \tag{7.16}$$

where a $*$ is used to denote a property calculated from the phase energy only, i.e. in the absence of interfacial energy $\left(\gamma_{\alpha\beta} = 0\right)$. From equation (5.10) we can see that the energy change $\Delta g_{\alpha\to\beta}$ becomes $\Delta g_{\alpha\to\beta} + \dfrac{2\gamma_{\alpha\beta}\Omega}{r}$ once the interfacial energy increase is accounted for. Hence the equilibrium condition of equation (7.16) becomes

$$\Delta\mu^{\alpha} = \Delta\mu^{\alpha*} + \frac{2\gamma_{\alpha\beta}\Omega}{\left(c_B^{\alpha*} - c_B^{\beta*}\right)r}.$$

Now if the solute is dilute, such that $c_B^{\alpha*} \ll 1$, the entropic term dominates the chemical potential of equation (3.9) so that $\Delta\mu^{\alpha} \approx RT\,\ln(c_B^{\alpha})$ and $\Delta\mu^{\alpha*} \approx RT\,\ln(c_B^{\alpha*})$.

The local concentration of solute B in the matrix α at the interface of a precipitate β of radius r is then

$$c_{\mathrm{B}}^{\alpha}(r) = c_{\mathrm{B}}^{\alpha*} \exp\left[\frac{2\gamma_{\alpha\beta}\Omega}{RT(c_{\mathrm{B}}^{\alpha*} - c_{\mathrm{B}}^{\beta*})r}\right] \tag{7.17}$$

where $c_{\mathrm{B}}^{\alpha*} = c_{\mathrm{B}}^{\alpha}(\infty) = \exp\left(\dfrac{\Delta\mu^{\alpha*}}{RT}\right)$ and $c_{\mathrm{B}}^{\beta*}$ are the equilibrium compositions of the solute for a flat interface, as calculated from the phase diagram.

Consider the case where $c_{\mathrm{B}}^{\beta*} < c_{\mathrm{B}}^{\alpha*}$, such that the precipitate ejects solute B as it grows. In this case $c_{\mathrm{B}}^{\beta*} - c_{\mathrm{B}}^{\alpha*} < 0$ and equation (7.17) tells us that $c_{\mathrm{B}}^{\alpha}(r) \leqslant c_{\mathrm{B}}^{\alpha*}$, i.e. the solute concentration around smaller particles is lower than that around larger particles. For example, using the parameter values for the calculations in figure 5.4, we see a decrease in the equilibrium matrix composition of 20% for a 1 nm particle, 2% for a 10 nm particle and 0.2% for a 100 nm particle. Hence we would expect a fine dispersion of nanoparticles to coarsen faster than a coarse dispersion of larger particles. A large interfacial energy also increases this effect and accelerates coarsening. This concurs with the predictions of equation (7.15). The Gibbs–Thompson effect, combined with growth, are further illustrated using a computational approach in section 8.1.

References

[1] Cocks A C F, Gill S P A and Pan J 1999 Modelling of microstructural evolution *Adv. Appl. Mech.* **36** 81–162

[2] Lifshitz I M and Slyozov V V 1961 The kinetics of precipitation from supersaturated solid solutions *J. Phys. Chem. Solids* **19** 35–50

Chapter 8

Modelling solid–solid phase transitions

In this section we are finally ready to put all aspects of the work covered in the preceding chapters into action and simulate the microstructural evolution of a multi-phase, multi-particle solid alloy system. Firstly, in section 8.1, a system of just two discrete particles is considered, to illustrate the different processes of growth and coarsening. In section 8.2 the evolution of a large number of particles is represented in a simpler fashion using a time-dependent particle size distribution. In section 8.3 this type of model is extended to cover massive transformations (where growing particles impinge upon one another), which has consequences for even simpler microstructural models that can be used in rapid, predictive software in section 8.4.

8.1 A two particle model

Consider the evolution of two spherical precipitate particles of radii, $r_1(t)$ and $r_2(t)$, over time within a total volume V_0, as shown in figure 8.1. We continue with the same thermodynamic example that has been employed throughout Part II of this book, namely the binary system (components A and B) for precipitation of a solid (S) phase in a liquid (L) phase with a composition of A of $c_B = 0.2$, as shown in figure 5.2(a). The temperature is 700 °C, generating an undercooling of 22 °C. Although this model concerns precipitation of a solid (β) phase in a solid (α) phase, the general energetic model is still acceptable. We determine the driving force for growth and coarsening, $\Delta g_{\alpha \to \beta}(c_M)$, at any instant from equation (5.6) based on the average matrix composition at the time, $c_M(t)$. At the start of the simulation, we have $c_M(0) = c_0$, which is the nominal composition of the material. As the particles grow, the average matrix composition increases as solute is ejected from the particles. Hence we use a (non-equilibrium) lever rule to determine the instantaneous matrix composition from

$$(1 - f)c_M + fc_\beta = c_0 \tag{8.1}$$

doi:10.1088/978-0-7503-3147-0ch8   8-1

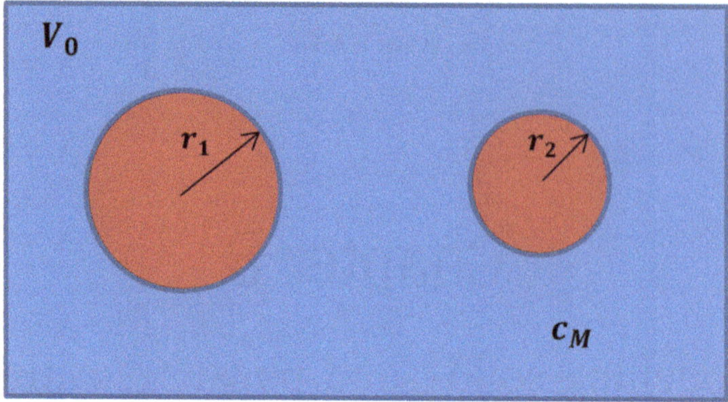

Figure 8.1. A simple two particle system. The microstructural evolution is defined by the change in the particles sizes, $r_1(t)$ and $r_2(t)$, over time.

where $f = \dfrac{\frac{4}{3}\pi(r_1^3 + r_2^3)}{V_0}$ is the volume fraction of the precipitating β phase.

The simulation starts by introducing two stable nuclei. Below the critical radius particles will quickly disappear. Hence the first particle (1) is introduced with a radius 10% above the critical radius r^*, and the second particle (2) is just 1% above the critical radius. They are therefore both expected to grow initially, but the first particle has a slight advantage over the second one. The radii of the two particles are evolved using equation (7.10) for growth and coarsening

$$\dot{r} = \frac{K}{r}\left[\frac{1}{r^*} - \frac{1}{r}\right]$$

where the critical radius is given by equation (5.11)

$$r^* = -\frac{2\gamma_{\alpha\beta}\Omega}{\Delta g_{\alpha\to\beta}(c_M)} \tag{8.2}$$

and the particles 'interact' through the evolution of the matrix composition $c_M(t)$.

The results of this simple simulation are shown in figure 8.2. A dimensionless measure of time is used such that $\bar{t} = \dfrac{K}{V_0}t$. In figure 8.2(a), the radius of the first particle is shown in red, and that of the second in blue. The simulation is not very representative, in the sense that the particles are in a very small volume, so that the coarsening happens quite quickly compared to the growth. Typically growth occurs relatively quickly (a few hours or days) and coarsening is very slow (occurring over months if not years, depending on the temperature). Initially both particles grow rapidly, although the first particle keeps its size advantage over the second. The larger particle then proceeds to consume the other particle during the coarsening phase, until the second particle completely disappears. Figure 8.2(b) shows the critical radius evolving over time. It increases as the driving force for growth, $\Delta g_{\alpha\to\beta}$,

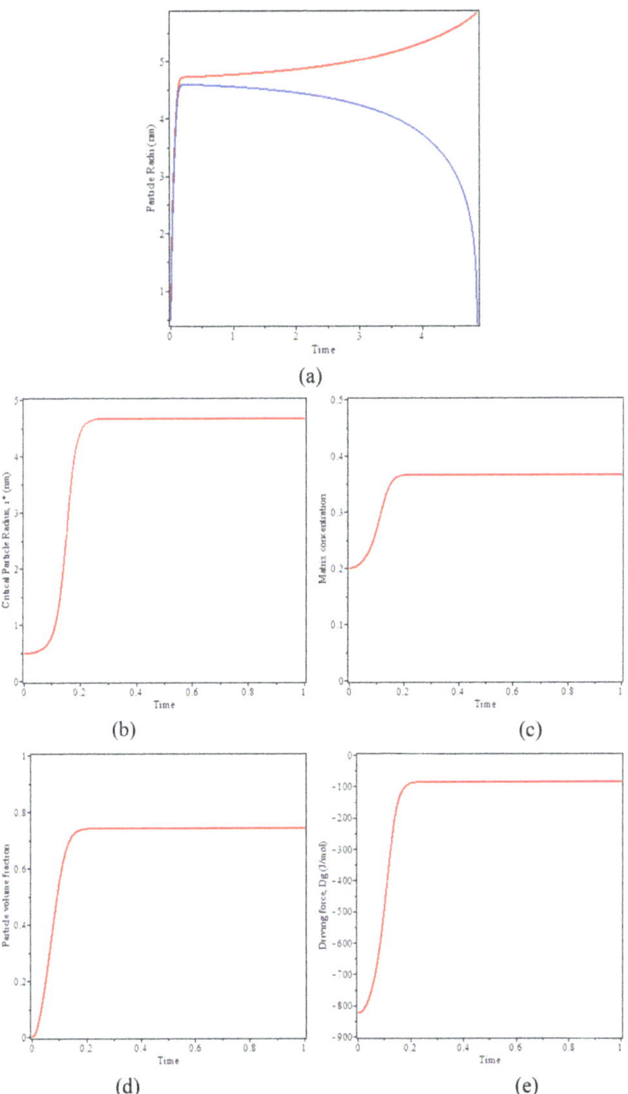

Figure 8.2. Results of the two particle model. (a) The particle radii, (b) the critical radius r^*, (c) the average matrix composition c_M, (d) the particle volume fraction f, (e) the driving force $\Delta g_{\alpha \to \beta}$.

decreases, as shown in figure 8.2(e). It reduces because the average matrix composition, c_M, increases (see figure 8.2(c)) as the total particle volume fraction grows (see figure 8.2(d)) and the particles eject solute into the matrix.

The onset of the coarsening phase is indicated by a plateauing of the variables in figures 8.2(b)–(e). At this point the critical radius becomes the average radius of the two particles. The driving force $\Delta g_{\alpha \to \beta}$ does not reach zero as the system is undercooled. In fact, none of these variables quite reach the equilibrium values predicted from the phase diagram. This is because the concentration around the two particles is lowered due to their interfacial energy.

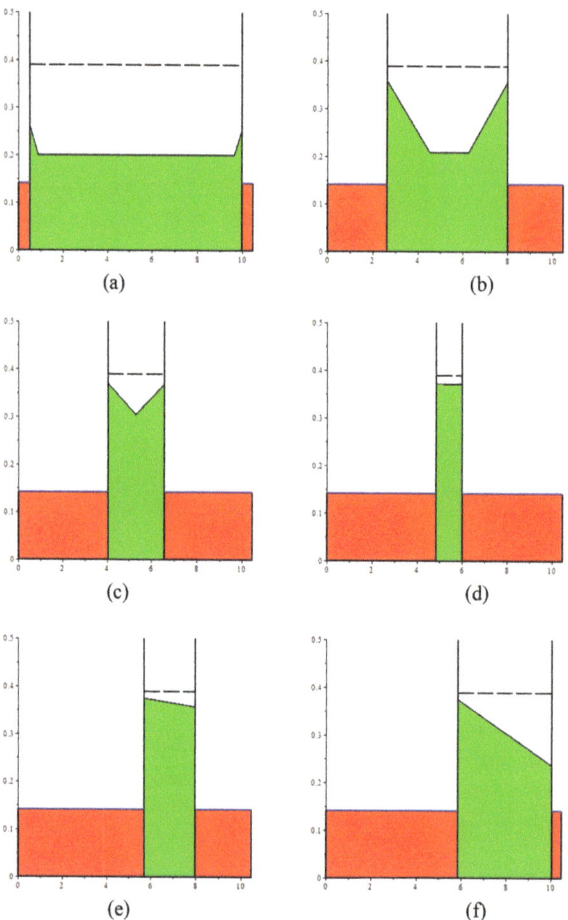

Figure 8.3. Snapshots of the concentration profiles within the two particle system at different times. The particle solute is shown in red, the matrix solute is shown in green. The dashed line indicates the equilibrium matrix composition, $c_B^{\alpha*}$. (a) initial set up, (b) the growth phase has started, (c) particles are on the verge of interacting with each other as their concentration fields touch, (d) growth is complete and particles have just started to exchange solute and coarsen, (e) the larger left particle starts to gain an advantage over the smaller right particle due to coarsening, (f) the left particle is on the verge of completely consuming the right particle.

The physics of the growth and coarsening phases are illustrated in figure 8.3, which shows a series of snapshots of the concentration profiles between the two particles (along a line joining their two centres) at different times. Note that the model does not calculate the concentration field, and does not take into account the distance between the particles, so these are determined on the simple volumetric basis using equation (8.1) with the local interfacial concentrations in the matrix determined using the Gibbs–Thompson relation of equation (7.17). The solute is shown in red in the particles and green in the matrix. The larger particle (1) is on the left and the smaller particle (2) is on the right, with the matrix between them. The horizontal dashed line is the equilibrium composition in the matrix, $c_B^{\alpha*}$, in the absence of interfacial effects. The solute in the precipitates is assumed to always adopt the equilibrium composition, $c_B^{\beta*}$. The starting configuration is shown in figure 8.3(a). The particles initially grow and eject solute in front of them, as shown in figure 8.3(b). This gradually slows the growth as the matrix composition increases. Eventually, in figure 8.3(c), the two ejected solute fronts meet and the solute starts to accumulate between them. When the solute distribution in the matrix is virtually uniform, as shown in figure 8.3(d), the coarsening phase begins. At this point the two particles start to interact, in the sense that their concentrations fields begin to overlap. The larger (left) particle has a slightly higher concentration at its interface due to the Gibbs–Thompson relationship. Initially this creates a small concentration gradient from the larger particle to the smaller one. Solute diffuses down the concentration gradient away from the larger particle into the smaller one. As solute ejection is required for particle growth, this allows the larger particle to grow at the expense of the smaller one. After a while, figure 8.3(e) shows that a substantial compositional gradient appears and the growth of the larger particle at the expense of the smaller one accelerates. Figure 8.3(f) shows the configuration just before the smaller (right-hand) particle disappears completely.

8.2 Population balance (or KWN) models

The previous section illustrated what is happening during the growth and coarsening of a simple system of two discrete particles. This could easily be extended to a more representative system with many thousands of particles. However, it is not efficient or necessary to simply extend the previous model to include so many particles. In this case it is more efficient to represent the state of the particles in the microstructure by a particle size distribution function. An example is shown in figure 8.4(a). Here particles of different sizes are placed into bins, where each bin represents a range of particle sizes. So for instance, bin i contains all particles with a radius between $r_{i-1} \leqslant r \leqslant r_i$, where $\Delta r_i = r_i - r_{i-1}$ is the width of the bin. In the case of figure 8.4(a), a uniform bin size is chosen, with the distribution divided into 20 bins for a range of particles up to 2 μm, such that $\Delta r = 0.1$ μm. The quantity of interest is the number of particles (per unit volume) within each bin, $N_i \Delta r_i$. For a given representation, this depends on the bin size. If the bin size is small, then there will be fewer particles within it, and if it is large then there will be more. In general an appropriate bin size is chosen to render the particle distribution function reasonably continuous, as in figure 8.4(a). The total number of particles (per unit volume) in the system

$$N_{\text{TOT}} = \sum_i N_i \Delta r_i$$

will not depend on the bin size, where the summation is over all bins. The particle size distribution is therefore similar to a probability density function. Other quantities can also be determined in this fashion, such as the particle volume fraction

$$f = \frac{4}{3}\pi \sum_i r_i^3 N_i \Delta r_i.$$

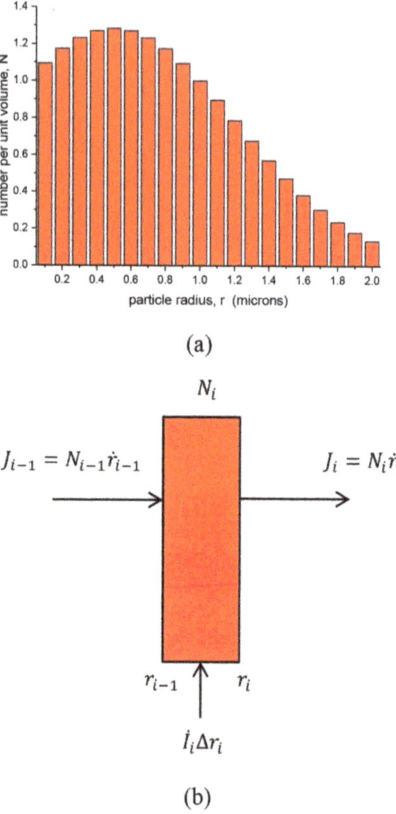

Figure 8.4. (a) A particle size distribution at a given time. (b) Particle fluxes in and out of bin i lead to a change in the number of particles (per unit volume), N_i, contained within it.

As the microstructure evolves, the particle distribution will evolve over time to reflect this. This type of model is known as a **population balance model** or, in the field of material science, a **Kampmann–Wagner Numerical (KWN) model**. Population balance means that the total number of particles is conserved, but particles can move from one bin to another, as they grow or shrink, hence changing the distribution over time. The exception to this is the first bin, containing the smallest particles. New particles can be introduced into this bin (due to nucleation) or removed (due to coarsening). As such, the total number of particles, N_{TOT}, can change over time, increasing due to nucleation, and decreasing due to coarsening.

The change of the particle number in each column, $\mathrm{d}N_i\Delta r_i$, depends on the fluxes of particles into and out of it. In figure 8.4(b), the particle number flux from bin i to bin $i + 1$ is $J_i = N_i\dot{r}_i$. This depends on the number of particles in the column and the rate at which their radius is changing. So if the average particle in bin i is growing, $\dot{r}_i > 0$, then these particles will be getting bigger and there will be a net flux of particles into bin $i + 1$. And vice versa, if the particles are shrinking then there will be a net flux into bin $i - 1$. There is an additional source flux, $\dot{I}_i\Delta r_i$, which is the rate at which the number of particles (per unit volume) are introduced into bin i, where $\dot{I}_i > 0$.

This represents new particles appearing through nucleation. Typically the critical nucleus size, r^*, falls within bin 2 (i.e. $r_1 \leqslant r^* \leqslant r_2$) such that the only non-zero contribution is I_2, which is defined by equation (7.1). Any particles that move from bin 2 into bin 1 ($0 \leqslant r \leqslant r_1$) are discarded as they have shrunk below the critical radius (due to coarsening), i.e. we keep $N_1 = 0$.

Population balance means the change in particle number in bin i over a time period Δt is given by $dN_i \Delta r_i = (J_{i-1} + I_i \Delta r_i - J_i)\Delta t$ which can be expressed more conveniently in discrete rate form as

$$\dot{N_i} = -\frac{(J_i - J_{i-1})}{(r_i - r_{i-1})} + I_i$$

or in continuous form, using $J = N\dot{r}$, we get the KWN model

$$\frac{\partial N}{\partial t} + \frac{\partial}{\partial r}\left(\frac{dr}{dt}N\right) = I \tag{8.3}$$

where $N(r,t)$ is a function of radius and time. The specific response of a material system is defined by the particle growth rate, $\dfrac{dr}{dt}$, from equation (7.10), and the nucleation rate, I, from equation (7.1). Both of these quantities depend on both the thermodynamics and the kinetics of the material and its processing conditions. It is also possible to apply this model to multiple phases, k, where a separate particle size distribution for each phase, $N_k(r, t)$, is introduced. In this case, the different distributions all evolve according to equation (8.3) but interact through their shared matrix composition, $c_M(t)$. The model also applies to multiple components, j, where the composition of each species within the matrix, c_{Mj}, is influential in determining the energetic driving force. In this case the kinetic coefficient for the key component with the lowest diffusion constant in a phase (typically a substitutional element) is used to determine the growth rate, e.g. for titanium carbide (TiC) the diffusion of the interstitial C is very fast compared to that of the substitutional Ti, so the latter controls the growth rate.

Equation (8.3) is readily implemented in a numerical code once the thermodynamics and kinetics are defined. PANDAT has a feature called *PanPrecipitation* to do this. In addition to the standard thermodynamic and kinetic data in the TDB file, additional information relating to the interfacial energy and nucleation rate needs to be provided in a Kinetic Database (KDB) file. The following instructions show how a KWN analysis of the Ni–Al system can be achieved using one of the included PANDAT examples.

PANDAT: conducting a KWN analysis of the Ni–Al system
  1. Create a 'New Workspace', select 'PanPrecipitation' and press 'Create'.
  2. Go to 'Database → Load TDB or PDB file' and locate the directory where PANDAT is installed, e.g. `C:\Program Files (x86)\CompuTherm LLC.`

3. Open the directory `Pandat 2020 Demo\Pandat 2020 Examples\PanPrecipitation\Ni-14%Al`.

4. It is possible to run the analysis automatically using `Batch Calc → Batch Run` and selecting the batch file `Ni-14AL.pbfx`, but the following step-by-step instructions are provided to give better understanding.

5. Open the file `AlNi_Prep.tdb`. This contains the phase energies and diffusivity data for the Ni–Al system. Choose 'Sel/Clr All' components and click 'OK'.

6. Go to 'PanPrecipitation → Load KDB or EKDB file' and select the file `Ni-14Al_Precipitation.kdb`. Click 'OK' to accept the settings.

7. Right click on the 'Ni-14Al-Precipitation' file listed in the 'Databases' window to 'Open with Text Editor' to see the contents of the file.

8. In the file you can see the additional quantities that are needed for the analysis, namely 'Molar Volume' (Ω—the volume of a phase is needed to work out its radius, area etc), 'Interfacial Energy' (γ—needed for coarsening),and 'Atomic Spacing' and 'Nucleation_Site_Parameter' (the fraction of nucleation sites available for nucleation) are needed to define the nucleation rate, I.

9. Go to 'PanPrecipitation → Precipitation Simulation' to start the simulation.

10. Set the Al concentration 'x(Al)' to '0.14' and the Ni concentration 'x(Ni)' to 0.86.

11. Go to 'Options' and check the 'Hour' checkbox under 'Time' units. Click 'OK'.

12. Note that the 'Thermal History' is set to run the simulation from 0 to 100 h, with the temperature set at 0 h and 100 h to 500 °C. This sets this to be a constant temperature over this period, as shown in the graph below. Note that these settings can be changed to model more complex heat treatments.

13. In the 'Intermediate PSD outputs' window in the bottom right corner, press the '+' button. The entry '2.0' will appear in the row below 'time (hr)'. This means that the particle size distribution after 2 h will be saved in the output. Press '+' three more times, and change the entries to '5', '10' and '50' hours for outputs at these times as well. Click 'OK'. The simulation will take a few seconds to complete.

The 'Default.graph' output shows the volume fraction f of the gamma-prime (γ') phase, also known as 'L12_FCC', evolving over time, with time on a base 10 log axis, as shown in figure 8.5(a). This is the primary strengthening phase in nickel-based superalloys, and hence is of great interest to someone developing such a material.

1. To plot the volume fraction as a linear function of time, right click on 'Table' in the file list in the left-hand 'Workspace' window and select 'Add a New table'.

2. Change the 'Table Name' from 'generated' to 'f vs time'.

3. In the vertical left-hand column are the variables available to plot. Hover the cursor over these to see what is available. Left click on 'time' and drag it into the first row of 'Columns' on the right whilst holding down the mouse button. It should appear in the first row of this table. Now do the same for 'vft' to place it in the second row. Click 'OK'.

4. A table with this data has been created. At this point it is possible to cut-and-paste this data into a separate plotting package if you wish, but to create a PANDAT plot, highlight both columns (hold down the CTRL button and click the header of the two columns) and then go to 'Table' on the upper option bar and select 'Create Graph'.

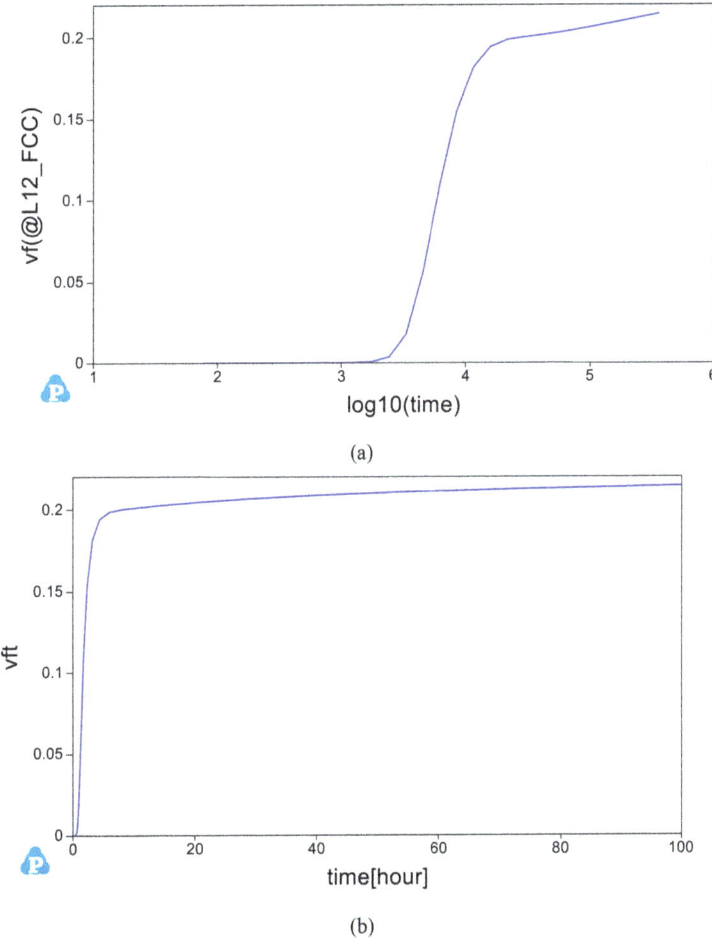

Figure 8.5. Volume fraction of L12_FCC (gamma prime) calculated as a function of (a) log(time) and (b) time. Created using PANDAT [2].

The resulting plot should look like figure 8.5(b). It can be seen that the gamma-prime phase precipitates out quite rapidly at 500 °C, reaching its equilibrium value within 5–10 h. Go back and try exploring the effect of different heat treatments if you wish, or try plotting other variables evolving over time, such as 'nd', the total number of particles per unit volume (equivalent to N_{TOT}) or see how changing the interfacial energy (in the KDB file) affects the result.

As encountered in the two particle model of section 8.1, the volume fraction quickly reaches equilibrium (or near it), but the particle sizes will still continue to evolve due to coarsening. The 'psd_100.graph' plot that also appears when you run this analysis shows the particle size distribution after 100 h. The following shows you how to plot the evolution of the particle size distribution at a few selected times over this period.

1. Double click on the table 'psd_2' in the left-hand 'Table' Workspace window to open it. Holding CTRL, click on the headers of the two columns 'psd_s (@L12_FCC)' (particle size) and 'psd_nd(@L12_FCC)' (particle number) to select them. Go to 'Table → Create Graph' in the upper horizontal selection bar.
2. Go to 'Graph → Edit Plots' on the upper horizontal bar. Note the current plot is defined in the 'New Plots' window. We wish to add the plots at other times to this window.
3. Firstly, click on the row in 'New Plots' with showing the current plot and press the 'X' button in the top right corner to delete it. This is deleted as PANDAT automatically rescales the first plot, whereas we wish to have the same scaling on all plots.
4. Click on 'Import a table file' and choose 'psd_2'. In 'Available Columns', drag 'psd_s(@L12_FCC)' and 'psd_nd(@L12_FCC)' into the 'New Plots' window in the same places as they were in the previous plot. Repeat this for 'psd_5', 'psd_10', 'psd_50' and 'psd_100' in the rows below and click 'OK'.
5. Click on the horizontal axis to rescale it from 0 to 3×10^{-9} (m) and the vertical axis from 0 to 1.7×10^{24} (number per m^3). You can also rename the axis titles and add a legend.

The resulting plot, shown in figure 8.6, is a number of snapshots of the number density of particles as a function of their size at different times. The first distribution, at 2 h, exhibits two peaks. The smaller left-hand peak is due to a lot of small particles being generated around 0.3 nm in size, the critical radius. This shows that nucleation is still continuing after 2 h, but many of the nuclei have grown to about 1 nm in size. The distribution after 5 h no longer exhibits two peaks, showing that nucleation has stopped. Nucleation stops naturally as the average matrix composition increases as solute is ejected from the growing particles, making further nucleation energetically unfavourable. The peak has shifted to the right, indicating that the average particle has increased in size since the 2 h mark. The peak continues to diminish and shift to the right at time progresses. This shows that the average particle size is getting bigger and the number of particles is necessarily getting smaller due to coarsening. If the evolution is extended beyond 100 h, the coarsening process will continue indefinitely, albeit at an increasingly slower rate, according to equation (7.15). This information about the particle size distribution is useful as it can be used to predict the strength of the alloy at different stages in its life.

These calculations are for a temperature of 500 °C. It is of interest to a metallurgist to see how a change in temperature affects the precipitation rate.

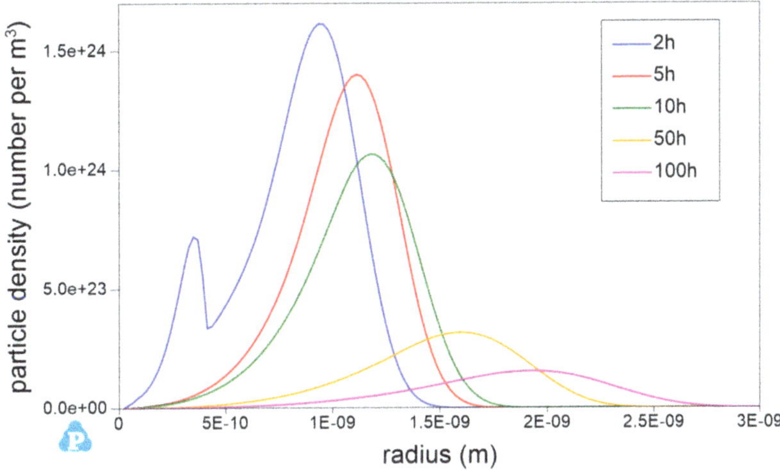

Figure 8.6. Particle size distribution plots for the gamma-prime phase in the Ni–Al system plotted at different times. Created using PanPrecipitation in PANDAT [2].

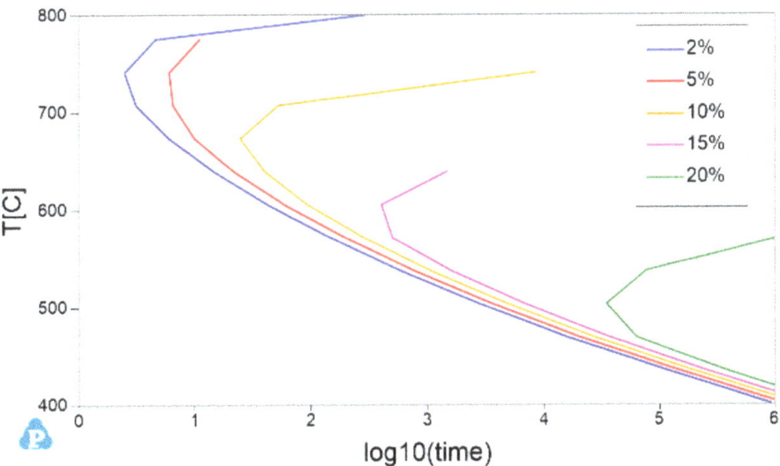

Figure 8.7. A TTT diagram showing the time for a specified volume fraction of the gamma-prime phase to precipitate in the Ni–Al system at different temperatures. Created using PANDAT [2].

The easiest way to do this is to plot a **Time–Temperature–Transformation (TTT)** diagram. In this plot, the time taken for a prescribed volume fraction of the precipitate phase to appear is plotted as a function of temperature. You can plot a TTT diagram in PANDAT by selecting 'PanPrecipitation → TTT Simulation'. Note that the default volume fraction is 2%. The resulting output is the blue curve shown in figure 8.7. Additional curves for other volume fractions have been added, in the same fashion as the additional PSD curves were added in figure 8.6. The curves exhibit the classical TTT shape, with an optimal temperature for rapid precipitation.

This is around 750 °C for a volume fraction of 2% and 500 °C for a volume fraction of 20%. The reason for this 'nose' shape is similar to that for the peak in nucleation rate, seen in figure 7.1(c), where slow kinetics reduce the precipitation rate at temperatures below the peak, and small undercooling (low thermodynamic driving force) reduces the precipitation rate above the peak. Each curve ends at an upper point as the temperature increases. This is because the specified volume fraction does not precipitate at temperatures above this point. Figure 8.7 shows that 500 °C is the optimal heat treatment for the precipitation of 20% volume fraction of gamma prime. Note, however, that if a component is to operate in service at an elevated temperature, the volume fraction will gradually transition to the equilibrium value for that temperature over a long time. Also note that each temperature on these curves requires an evolution of the PSD to be simulated which takes a considerable amount of computation time. A faster approximate method is introduced in section 8.4.

8.3 Massive transformations

Massive transformations are phase transformations in which the volume fraction of the new phase is large, typically greater than 50%. Examples of massive transformations are solidification, when a 100% liquid phase changes to a 100% solid phase, and the martensitic transformation in steels, when 100% austenite change to 90% or more martensite (there is typically 5%–10% retained austenite remaining). The principal consequence of massive transformations is that particles will make contact with each other as they grow. This eventuality has not been accounted for in previous analyses in this chapter, but can be accommodated in a simple fashion using the Avrami equation.

The **Avrami equation** is also known as the **Johnson–Mehl–Avrami–Kolmogorov (JMAK) equation** as it was developed independently by a number of scientists. It accounts for the overlap of particles at high volume fractions. It does not model the contact of particles explicitly, but allows particles to continue to evolve independently and overlap each other, as shown in figure 8.8(a). The Avrami equation takes the overlapping regions into account, so that they are not double counted (or more). This introduces the concept of an **extended volume fraction**, f_{ext}. This is the predicted volume fraction of the system if the overlapping of particles is not taken into account. As such, the extended volume fraction can be greater than one. The extended volume of the particles in figure 8.8(a) is the sum of the volume of all the spheres shown. The aim is to determine the actual volume fraction, f, which does not include any overlaps, as shown in figure 8.8(b).

Derivation of Avrami equation

The Avrami model postulates that a small increase in the actual volume fraction, df, must be proportional to the product of the small increase in the extended volume fraction, df_{ext}, and the amount of material that remains untransformed, $(1 - f)$

$$df = (1 - f)df_{ext}.$$

The transformed volume fraction is therefore determined from adding up these small increments such that

$$\int_0^f \frac{\mathrm{d}f}{1-f}\,\mathrm{d}f = \int_0^{f_{ext}} \mathrm{d}f_{ext}$$

which gives

$$-\ln(1-f) = f_{ext}$$

or, in final form, the **Avrami equation** is

$$f = 1 - \exp(-f_{ext}). \tag{8.4}$$

This function is the blue curve in figure 8.9. It shows that about 90% of the volume is expected to have transformed by the time the extended volume fraction reaches 250%. The red line shows the extended volume fraction for comparison. It is almost the same as the actual volume fraction for transformations involving less than about 20% of a system. This is because there is little chance of overlap at these small volume fractions. Hence the Avrami equation can always be applied to a phase transformation, but for

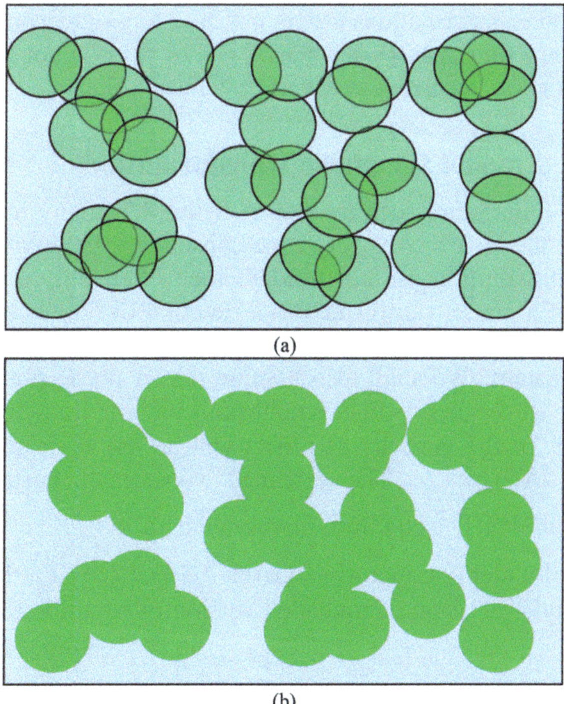

Figure 8.8. If spherical particles grow to a large volume fraction they will eventually overlap. (a) shows the extended volume fraction, f_{ext}, of particles (including overlaps) and (b) shows the actual volume fraction, f, transformed.

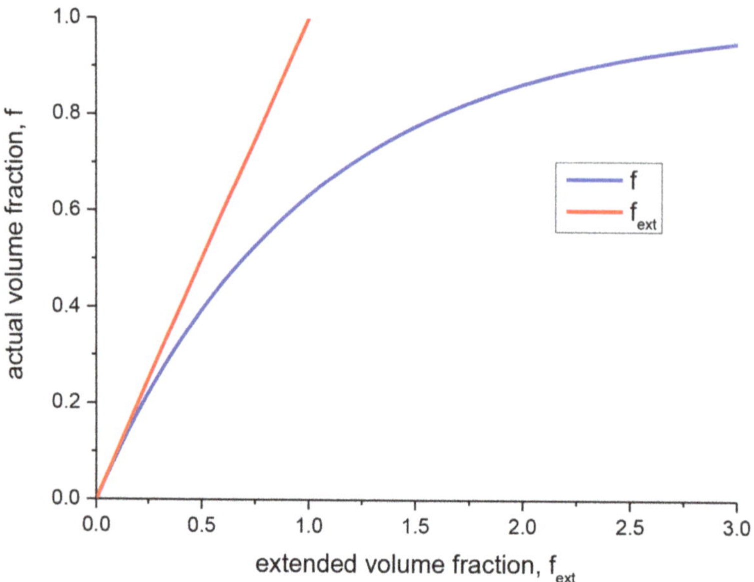

Figure 8.9. The Avrami model predicts the actual volume fraction, f, (not counting particle overlaps) as a function of the extended volume fraction, f_{ext} (counting particle overlaps).

low volume fraction transformations it does not have any significant effect. Note that the volume fraction f only reaches 1 when $f_{ext} \to \infty$. This is not the case, however, with full transformation occurring at a finite extended volume.

8.4 A simplified model for phase transformations

To produce a TTT diagram, such as that in figure 8.7, many hundreds of KWN simulations are required, which is very time consuming. However, if details about the particle size distribution are not needed, then often only a prediction of the volume fraction of a phase transformed as a function of time is sufficient. For this simplified approach, let us assume that the nucleation rate I and growth rate of the particles r are constant. We shall also assume that a phase proceeds through its growth phase until it reaches its final equilibrium volume fraction, f_{eq}. In this case the average radius of the particles at time t will be $r(t) = rt$, and the number of particles will be $N(t) = It$. The extended volume of particles is therefore $V_{ext} = \frac{4}{3}\pi r^3 N = \frac{4}{3}\pi \dot{r}^3 I t^4$. The equilibrium volume is then $f_{eq} V_0$, where V_0 is the total volume of material under consideration. Given $f_{ext} = V_{ext}/f_{eq} V_0$, the transformation of the phase is modelled using an amended version of equation (8.4) as

$$f = f_{eq}[1 - \exp(-\alpha t^n)] \tag{8.5}$$

such that $f \to f_{eq}$ as $t \to \infty$, where in this case $\alpha = \dfrac{4\pi \dot{r}^3 I}{3 f_{eq} V_0}$ and $n = 4$. Different scaling relationships can be derived for non-spherical phases, or non-constant nucleation

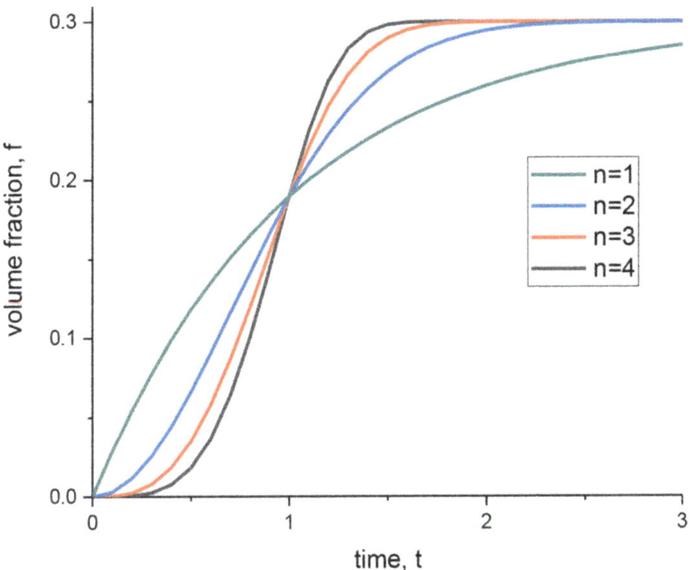

Figure 8.10. The effect of the exponent n on the temporal evolution of volume fraction, f, according to equation (8.5).

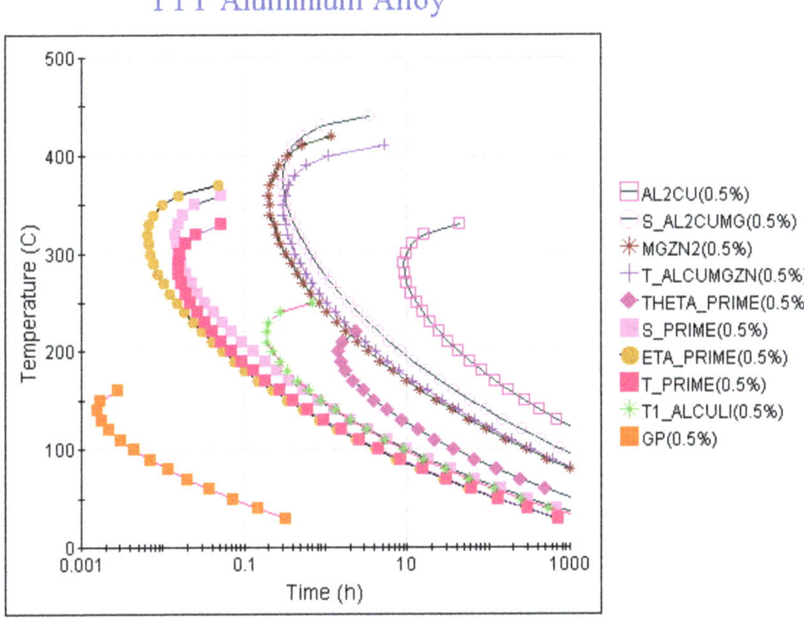

Figure 8.11. A TTT diagram for an Al alloy created using the software JMatPro® [1]. The times for the precipitation of 0.5% volume fraction are shown for 10 different phases as a function of temperature.

and growth rates, e.g. $n = 3$ for disk-shaped grain boundary phases. Figure 8.10 shows the effect of different n exponents on the model of equation (8.5) for a value of $f_{eq} = 0.3$.

The software JMatPro [1] uses this approach for modelling some material systems (but not steels). This combines thermodynamic and kinetic information into a simplified expression of the form of equation (8.5), which allows TTT diagrams to be plotted very rapidly. An example generated by JMatPro for an Al alloy is shown in figure 8.11. This is for a complex alloy with 8 different components. The TTT diagram changes if the composition of the alloy is changed, based on the thermodynamics and kinetics for the composition. There are 10 different phases that appear over the range of temperatures. Based on diagrams such as this, the temperature history of a material can be regulated to promote phases that add to the strength of the material and help to avoid deleterious (generally brittle) phases that can reduce its strength.

References

[1] JMatPro®, a software developed by Sente Software Ltd (UK) www.sentesoftware.co.uk/jmatpro [accessed 1 October 2021]

[2] Pandat: software suite for thermodynamic calculation and kinetic simulation of multi-component alloys. CompuTherm LLC, Madison, Wisconsin, USA www.computherm.com [accessed 1 October 2021]

Thermodynamics, Kinetics and Microstructure Modelling

Simon P A Gill

Chapter 9

Modelling liquid–solid phase transitions

Solidification is of immense practical importance, as nearly all manufacturing processes for metals and alloys involve casting them from a liquid. The nature of the liquid–solid transformation is different to the solid–solid transformations considered in chapter 8 for a number of reasons: (1) the transformation is typically much faster, (2) the latent heat release is much more substantial and influential, and (3) the liquid undergoes complete (massive) phase transformation. A consequence of this is that the liquid–solid interface rarely remains flat resulting in complex morphologies that can only be represented in a numerical simulation. This topic is investigated in sections 9.2 and 9.3, but before that there is a quick overview of simplified analytical solidification models. For more details, the reader is referred to [1, 2].

9.1 Simple solidification models for a binary system

These analyses are illustrated employing the liquid–solid thermodynamic model of the eutectic AB phase diagram of figure 3.11(c) previously employed for other examples in the book. The relevant region of the phase diagram is reproduced in figure 9.1. The nominal composition of the alloy is taken as $c_0 = 0.1$, at a composition before the solidus is truncated by the eutectic line. In this case, solid first appears at $T_1 = 762\,°C$ and finishes at $T_2 = 668\,°C$. To make analytical progress, the models in this section assume the liquidus and solidus lines are straight, such that

$$T_L = T_A + m_L c_L \qquad T_S = T_A + m_S c_S \tag{9.1}$$

where $m_L = -327\,°C$ and $m_S - 2235\,°C$ are the slopes of the lines, from table 3.1, and $T_A = 800\,°C$ is the melting temperature of pure A. These approximations are shown as red lines in figure 9.1. The approximation for the liquid is good, but the real solidus line curves differ from the approximation significantly. Both the real liquidus and solidus are also truncated by the eutectic line at $T_E = 656\,°C$, which is not accounted for in (9.1). The error introduced by these assumptions is assessed in

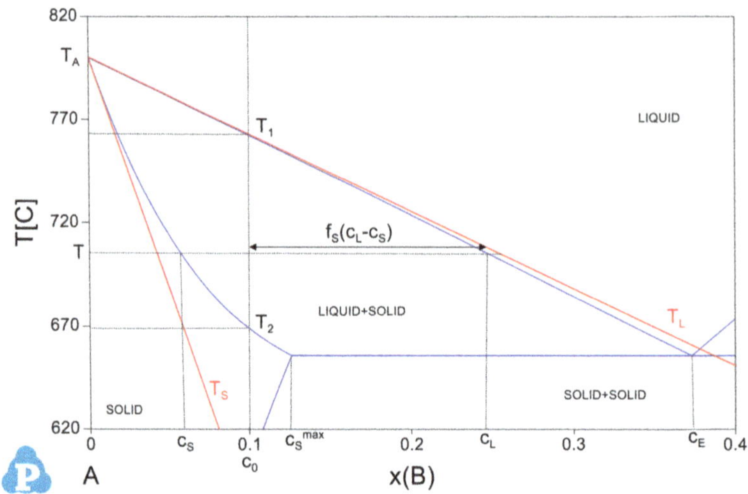

Figure 9.1. The binary phase diagram of figure 3.11(c). For a nominal composition of $c_0 = 0.1$, solidification starts at $T_1 = 762\,°C$ and finishes at $T_2 = 668\,°C$. The solid volume fraction, f_S, at a temperature T is calculated according to levers rule, equation (3.19). The liquidus and solidus are approximated by the two straight red lines T_L and T_S respectively. Created using PANDAT [7].

the following sections. The solidification process defined by equation (9.1) defines a constant partition coefficient, k. Following equations (3.20) and (3.13) this can be written as

$$k = \frac{m_L}{m_S} = \frac{c_S}{c_L} = \frac{T_A - T_1}{T_A - T_2} \tag{9.2}$$

as introduced in section 3.2.5, where $k = 0.17$ for the system in figure 9.1.

9.1.1 Global equilibrium

This is the simplest model for predicting the solid fraction, f_S, as a function of temperature, T. This uses levers rule (section 3.2.3) to predict the solid fraction. This essentially assumes that redistribution of solute within both phases is infinitely fast, either by diffusion or convectional/mechanical mixing, such that no concentration gradients exist in either phase. Example outputs of this type, produced by PANDAT, were previously encountered in figures 3.10 and 3.13. Equation (3.19) can be written in the above notation as

$$f_S = \frac{c_L - c_0}{c_L - c_S} = \frac{1}{1 - k}\left(1 - \frac{c_0}{c_L}\right). \tag{9.3}$$

Substituting equations (9.1) and (9.2) gives

$$f_S = \frac{1}{1 - k}\left(\frac{T - T_1}{T - T_A}\right). \tag{9.4}$$

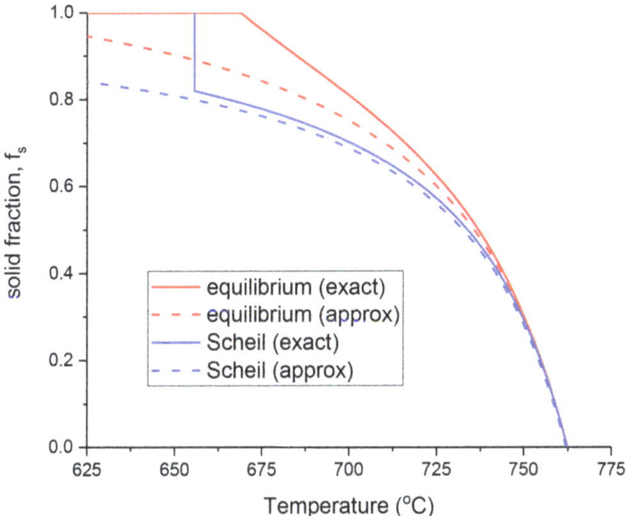

Figure 9.2. Solid fraction, f_s, calculated for the alloy in figure 9.1. The exact solutions were calculated using PANDAT [7], with the equilibrium and Scheil approximations given by equations (9.4) and (9.5) respectively.

The predictions of this model (red dashed line) is compared with the exact equilibrium model solution calculated by PANDAT (red solid line) in figure 9.2. It can be seen that the approximation underestimates the solid fraction at lower temperatures. This is because the straight line approximation to the solidus (T_S) in figure 9.2 predicts T_2 to be 585 °C whereas it is actually 668 °C.

9.1.2 Scheil equation

This is also known as the non-equilibrium lever rule. This applies where solute redistribution in the solid is slow (taken to be zero) compared to that in the liquid phase (taken to be infinitely fast). This means that concentration gradients can exist in the solid but not in the liquid.

Derivation of the Scheil equation

Consider the movement of a liquid–solid transformation front over a specimen of length L, as shown in figure 9.3(a). The position of the interface is related to the volume fraction such that $x = f_S L$. The solute concentration in the solid is frozen as the interface moves, so $c_S(x)$ varies within the solid. At the start of the transformation we have $c_L(0) = c_0$ and $c_S(0) = kc_0$. As the solid grows, solute is ejected into the liquid, where it is assumed to quickly redistribute to keep the concentration uniform. Now consider the movement of solute as the interface moves from x to $x + dx$, as shown in figure 9.3(a). Mass conservation requires that the two yellow shaded regions have the same area, i.e. solute ejected from the solid, $(c_L - c_S)dx$, must increase the solute concentration in the remaining liquid region by $dc_L(L - x)$. Equating these two gives

$$(c_L - c_s)Ldf_s = dc_L(L - Lf_S).$$

Given $c_S = kc_L$ and integrating both sides gives

$$(1 - k) \int_0^{f_S} \frac{\mathrm{d}f_S}{1 - f_S} = \int_{c_0}^{c_L} \frac{\mathrm{d}c_L}{c_L}$$

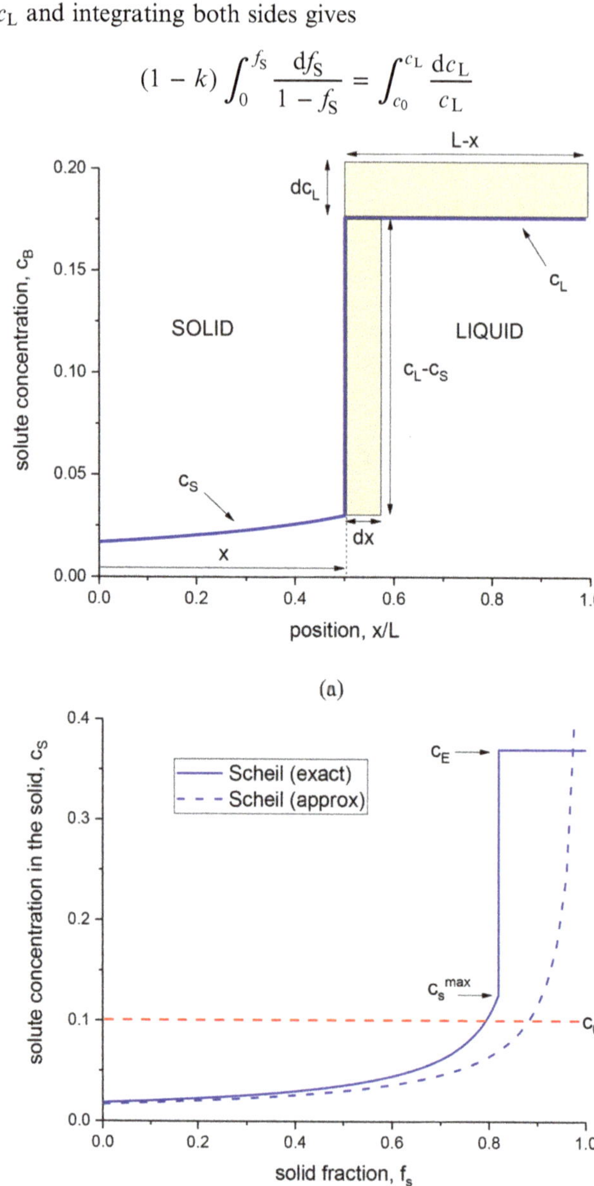

Figure 9.3. The Scheil model for solidification assumes rapid solute redistribution in the liquid and no redistribution in the solid. (a) A snapshot of the compositional profile at a given time. The shaded regions have equal area and show that the solute ejected from the solid as the interface advances must be redistributed within the remaining liquid. (b) The final Scheil compositional profile in the solid for the exact solution (calculated using PANDAT [7]) and the approximation of equation (9.5).

which results in the **Scheil equation**

$$c_S = kc_L = kc_0\left(1 - f_S\right)^{k-1}. \tag{9.5}$$

The predictions of equation (9.5) are compared with the exact Scheil solution calculated using PANDAT in figure 9.3(b). To conserve the total amount of solute, the area under each curve must be the nominal composition c_0. Equation (9.5) erroneously predicts that $c_S \to \infty$ as $f_S \to 1$ during the final stages of solidification, and in this case 50% of the solute is in the region of $0.99 \leqslant f_S \leqslant 1$. This cannot be true of course, and the exact solution indicates why. In figure 9.1 it can be seen that the real solidus and liquidus do not extend forever as they encounter the eutectic line, whereas the approximation of equation (9.1) has no such termination. In reality, the solute concentration in the solid reaches a maximum value, $c_S^{max} = 0.124$, at the eutectic temperature of $T_E = 656\,^\circ\text{C}$, as shown in figure 9.1. At this point a eutectic transformation occurs, where the solute concentration is $c_E = 0.37$. The predictions for the solid fraction as a function of temperature from equation (9.5) with equation (9.1) are compared with the exact solution from PANDAT in figure 9.2. The vertical line in the exact Scheil model shows the occurrence of the eutectic transformation in the solute rich liquid towards the end of the solidification process. The approximate solution is reasonable up to the onset of the eutectic transformation.

9.1.3 Steady state growth

As in the Scheil model, is it assumed that there is no solute redistribution in the solid. However, in this case there is redistribution within the liquid at a finite rate determined by the diffusion constant for solute in the liquid, D_L, i.e. concentration gradients are now possible within the solid and the liquid. The solid and liquid compositions begin at $c_S = kc_0$ and $c_L = c_0$. As the solid grows, solute is ejected and piles up in front of the moving interface. This increases the interfacial concentration of the liquid, as it can only diffuse away at a finite rate, and therefore the solute concentration in the solid also increases, in accordance with equation (9.2). After this initial transient, shown in figure 9.4(a), the solid concentration reaches $c_S = c_0$. At this point the interface can move forward without further solute ejection as local equilibrium is achieved at the interface such that $c_L = c_0/k$. The solute that was previously ejected still resides in front of the interface and must be moved along as the transformation proceeds. However, the concentration profile ahead of the interface does not change shape. This shape can be determined if we take the distance from the interface to be X, as shown in figure 9.4(a).

Derivation of steady state concentration profile
The solute concentration in the liquid changes due to diffusion, according to equation (6.15), as

$$\frac{\partial c_L}{\partial t} = D_L \frac{\partial^2 c_L}{\partial X^2}. \tag{9.6}$$

However the composition at a point also changes due to the movement of the coordinate system X with the interface. Given the composition $c_L(X)$ at position X, if the interface moves forward a small amount dX then the composition at the new position X is now $c_L(X + dX) \approx c_L(X) + \dfrac{\partial c_L}{\partial X} dX$. If the profile moves at velocity v then the movement over time dt is $dX = v \cdot dt$ and the contribution to concentration change is

$$\frac{\partial c_L}{\partial t} = \frac{c_L(X + dX) - c_L(X)}{dt} = v\frac{\partial c_L}{\partial X}. \tag{9.7}$$

In the steady state the rate of change of concentration must be zero. In this case, the two contributions from equations (9.6) and (9.7) cancel such that

$$D_L\frac{\partial^2 c_L}{\partial X^2} + v\frac{\partial c_L}{\partial X} = 0.$$

Integrating once, given $\dfrac{\partial c_L}{\partial X} = 0$ when $c_L = c_0$, gives

$$\frac{\partial c_L}{\partial X} = \frac{v(c_0 - c_L)}{D_L}.$$

Rearranging and integrating again, and setting $c_L(0) = \dfrac{c_0}{k}$ at the interface, gives

$$\int_{c_0/k}^{c_L} \frac{dc_L}{c_0 - c_L} = \frac{v}{D_L}\int_0^X dX.$$

The steady state profile of the solute concentration ahead of the liquid is therefore

$$c_L(X) = c_0\left[1 + \left(\frac{1 - k}{k}\right)\exp\left(-\frac{vX}{D_L}\right)\right]. \tag{9.8}$$

This profile is plotted in figure 9.4(a). The width of the profile is determined by the characteristic length D_L/v. The steady state profile will continue through the liquid until the outer limits of the diffusion front reaches the end of the sample. At this point the solute will start to accumulate in the liquid, forming a final transient. A schematic example of the final solute concentration across the sample is shown in figure 9.4(b). For mass conservation, the area under the curve must equal the nominal composition c_0L, so the area of solute deficit (below c_0) in the initial transient is equal to the area of the solute surplus (above c_0) in the final transient. The maximum concentration in the final transient is the eutectic composition, c_E.

The models in this section have assumed planar growth, i.e. a flat solid-liquid interface. This is rarely the case however, and to explore this further numerical simulation of the interface geometry evolving over time is considered. This is implemented through the phase field method, introduced in section 9.2. The effects of material and processing parameters on the interface stability are then investigated in section 9.3.

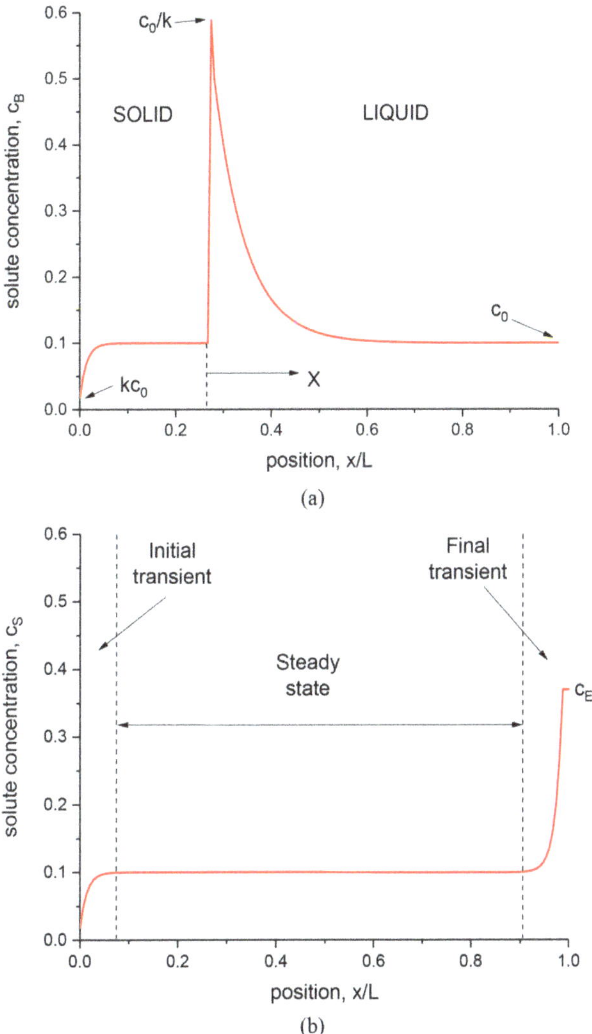

Figure 9.4. Steady state growth can occur when there is no solute redistribution in the solid, and solute redistribution in the liquid is limited by diffusion, allowing the solute composition at the interface to peak locally. (a) after an initial transient, a steady state concentration profile, defined by equation (9.8), moves ahead of the growing solid. (b) the final solute distribution in the solid is uniform except for a deficit due to the initial transient and an equal but opposite surplus due to the final transient.

9.2 A phase field model for solidification

A common problem when modelling large scale phase transformations is following the evolution of the interface. The growing phase can have a complex geometry, especially when phases impinge upon one another. The phase field model provides a solution to this problem by not explicitly tracking the interface. It allows the geometry to evolve naturally without any direct updating of the interfacial position or change in the underlying numerical grid. Here it is used to examine the stability of

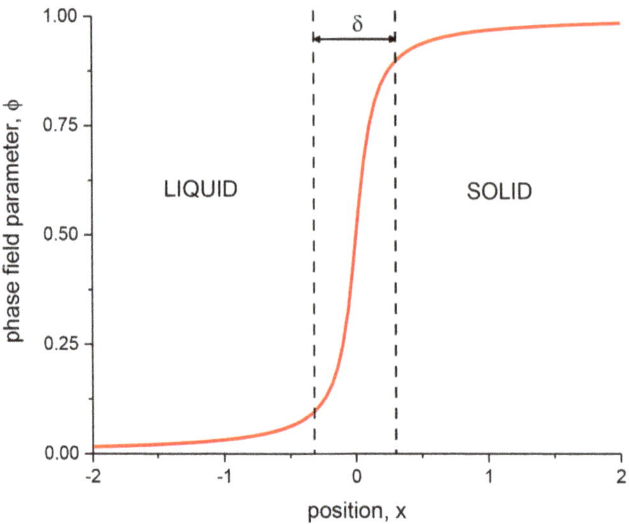

Figure 9.5. Schematic of the liquid–solid interface representation in the phase field method.

the liquid–solid interface during solidification. The material state at a point is defined by the phase field variable, $\phi(x,y,t)$. This can vary between zero and one, where $\phi = 0$ represents the liquid phase and $\phi = 1$ represents the solid phase. A typical phase field profile across an interface is shown in figure 9.5. The phase field method is known as a **diffuse interface** method, in that the phase change occurs smoothly over a finite distance, known as the interface width, δ. A simple implementation of the model for a pure metal (single component) system follows. See Tagaki [3] for extension of the model to binary and higher order systems and [4] for a review of the phase field method applied to other aspect of microstructural evolution in material systems.

The phase field parameter evolves to minimise the energy of the system

$$F(\phi, \nabla\phi) = \int_V f_{\text{phase}} + f_{\text{well}} + f_{\text{int}} \, dV \tag{9.9}$$

where here we follow Kobayashi [5] and take the phase energy to be a simple linear weighted sum

$$f_{\text{phase}} = \phi g_{\text{s}} + (1 - \phi)g_{\text{L}} \tag{9.10}$$

where g_{s} and g_{L} are the Gibbs free energies per unit volume of the solid and liquid phases respectively. The double well term

$$f_{\text{well}} = W\phi(1 - \phi) \tag{9.11}$$

is a general feature of the phase field method. It has minima of $f_{\text{well}} = 0$ at $\phi = 0$ and $\phi = 1$ and a maxima of $f_{\text{well}} = \frac{1}{4}W$ at $\phi = \frac{1}{2}$. The purpose of this term is to ensure that the phase field parameter prefers to sit in one of these minima. This ensures that, away from the interface, there is a single phase at each point ($\phi = 0$ or 1), rather than a uniform mix of both. The constant W defines the depth of the well and the energy barrier to phase transition. The interfacial term

$$f_{\text{int}} = \frac{1}{2}\varepsilon^2 \,|\nabla\phi|^2 \tag{9.12}$$

generates an energetic penalty where the phase field gradient, $\nabla\phi = \left[\dfrac{\partial\phi}{\partial x}, \dfrac{\partial\phi}{\partial y}\right]$, is non-zero. From figure 9.5, it is clear that the phase field gradient is only non-zero in the interface region, and hence this term represents the effect of an interfacial energy. Together the parameters ε and W define the interfacial energy γ and the interface width δ [3]. In the first instance, we assume that $\varepsilon = \varepsilon_0$ is a constant.

As seen elsewhere, such as equation (6.12), to minimise the energy of a system a state variable moves down an energy gradient that it can change such that

$$\frac{\partial\phi}{\partial t} = -M\frac{\partial F}{\partial\phi} \tag{9.13}$$

where the parameter M defines the mobility of the interface. Note that ϕ is not a conserved variable, i.e. the total amount of each phase can change. As such, unlike concentration say, the variable ϕ does not have to obey a mass balance relationship. Now for a constant (isotropic) interfacial energy $\varepsilon = \varepsilon_0$ equations (9.9)–(9.13) give

$$\frac{\partial\phi}{\partial t} = M\left[W\phi(1-\phi)\left(\phi - \frac{1}{2} - \frac{\Delta g(T)}{W}\right) + \varepsilon_0^2\nabla^2\phi\right] \tag{9.14}$$

where $\Delta g(T) = g_s - g_L$. For a pure metal, equation (5.1) gives

$$\Delta g(T) = -\frac{L_f(T_M - T)}{T_M}$$

where T_M is the liquidus temperature, $T_M - T$ is the undercooling and L_f is the latent heat of fusion in J m^{-3}. For stability purposes, it is necessary that the phase field variable satisfy $0 \leqslant \phi \leqslant 1$ which requires that $\left|\phi - \dfrac{1}{2} - \dfrac{\Delta g(T)}{W}\right| \leqslant 1$ in equation (9.14). This is satisfied if $\left|\dfrac{\Delta g(T)}{W}\right| \leqslant \dfrac{1}{2}$. To ensure that this is the case, Kobayashi [5] capped the undercooling to stay within this range by substituting the following expression

$$\frac{\Delta g(T)}{W} = -\frac{\alpha}{\pi}\tan^{-1}\left(\frac{L_f(T_M - T)}{T_M}\right) \tag{9.15}$$

where $\alpha < 1$ to keep this in the acceptable range.

Now equation (9.14) is solved in tandem with the temperature field, given by the standard heat conduction equation

$$\rho c_{\mathrm{p}}\frac{\partial T}{\partial t} = \nabla(k\nabla T) + \dot{Q} \tag{9.16}$$

where ρ, c_{p} and k are density, specific heat capacity and thermal conductivity respectively. For the purposes of this model, they are assumed to be the same in both phases. A heat source term, $\dot{Q} = L_{\mathrm{f}}\dot{\phi}$, is added to model the latent heat release at the interface during the transformation. It is easy to show that the total heat emitted (per unit volume) is $Q_{\mathrm{TOT}} = \int \dot{Q}\mathrm{d}t = L_{\mathrm{f}}\int \dot{\phi}\mathrm{d}t = L_{\mathrm{f}}\int_{0}^{1}\mathrm{d}\phi = L_{\mathrm{f}}$ as expected.

In the following simulations we are only interested in demonstrating the effect of the latent heat on the stability of the liquid–solid interface, rather than simulating a particular material system. Hence we follow Kobayashi [5] and define a general system with $\alpha = 0.9$, $W = 1$, $L_{\mathrm{f}} = 10$, $T_{\mathrm{M}} = 1$, $M = 3333$ and $\varepsilon_0 = 0.01$ in equation (9.14). We take $\dfrac{k}{\rho c_{\mathrm{p}}} = 1$ in equation (9.16) and write it in terms of the dimensionless temperature $\overline{T} = T/T_{\mathrm{M}}$ such that

$$\frac{\partial \overline{T}}{\partial t} = \nabla^2 \overline{T} + K\frac{\partial \phi}{\partial t} \tag{9.17}$$

where $K = \dfrac{L_{\mathrm{f}}}{\rho c_p T_{\mathrm{M}}}$ is the parameter of interest. The effect of changing this parameter is investigated in the next section.

9.3 Stability of the liquid–solid interface

Equation (9.14) with (9.15) and (9.17) are solved using the finite element package COMSOL Multiphysics v5.6. The geometry is 9 by 9 units square, with an unstructured triangular mesh of linear elements with side lengths of 0.03. An initial solid zone is introduced ($\phi = 1$, $T = 1$ for $x < 2$) on the left with the remainder of the geometry representing the liquid ($\phi = 0$, $T = 0$ for $x \geqslant 2$). All the external boundaries are thermally insulated. Snapshots of the phase field variable at different times are shown in figure 9.6, where the solid is shown as red and the liquid as blue. As the solid grows the interface moves from left to right over time. It can be seen that the interface remains flat for $K = 0.8$ but becomes increasingly unstable as K increases above 1. A similar result can be seen in figure 9.7, where a small nucleus of solid is introduced into the centre of the simulation to induce radial growth.

Before continuing, it is of interest to note that whilst the structures in figures 9.6 and 9.7 nicely demonstrate the effect of the latent heat measure K on the stability of the liquid–solid interface, they do not look like the structures typically seen in practice. This is because crystalline solids have a strong preference for forming flat interfaces (facets) in particular orientations, depending on their crystal structure. This phenomenon can be represented in the phase field model by expressing the interfacial energy measure, $\varepsilon(\theta)$, as a function of the angular orientation of the interface, θ. Many crystals have a hexagonal structure which favours six-fold symmetry (in 2D) such that a reasonable functional form is

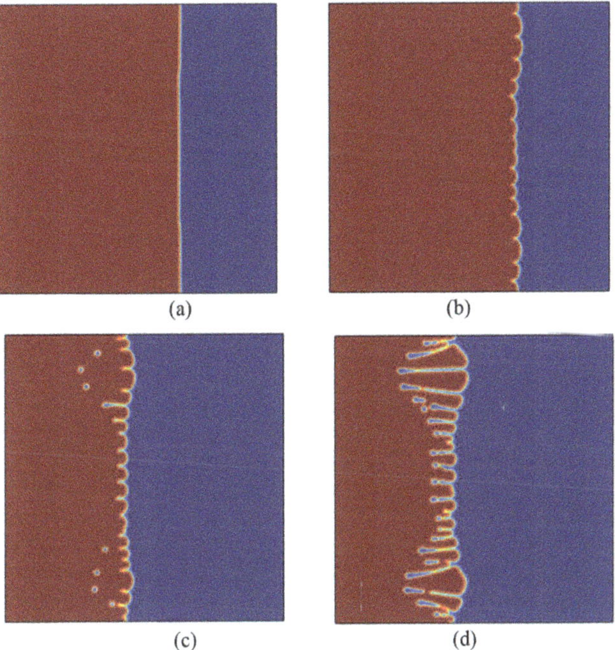

Figure 9.6. Phase field variable ϕ of a liquid (blue) to solid (red) phase transformation using equations (9.14) and (9.17) for an isotropic interfacial energy. Growth is from an initially planar solid region on the left. The different latent heats are : (a) $K = 0.8$, (b) $K = 1.0$, (c) $K = 1.1$ and (d) $K = 1.2$. Modelling performed using COMSOL Multiphysics® [8].

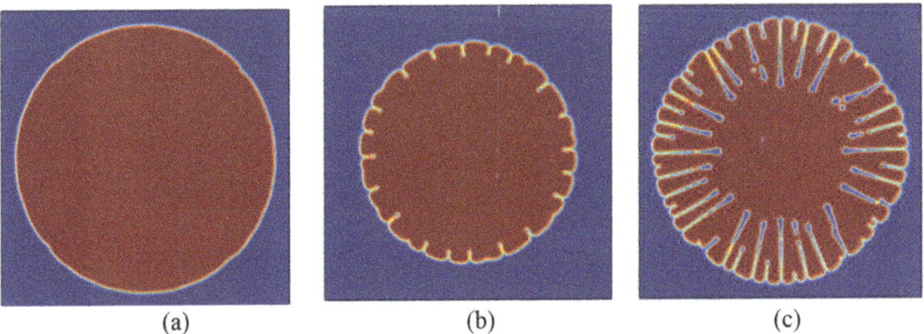

Figure 9.7. Phase field variable ϕ of a liquid (blue) to solid (red) phase transformation using equations (9.14) and (9.17) for an isotropic interfacial energy. Growth is from an initially circular solid seed region in the middle. The different latent heats are : (a) $K = 1.0$, (b) $K = 1.1$ and (c) $K = 1.2$. Modelling performed using COMSOL Multiphysics® [8].

$$\varepsilon(\theta) = \varepsilon_0(1 + \delta \cos(6\theta)) \tag{9.18}$$

where $\delta = 0.05$ is the amplitude of the interfacial energy variation with orientation, and $\tan \theta = \dfrac{\partial \phi / \partial y}{\partial \phi / \partial x}$, i.e. $\theta = \theta(\nabla \phi)$. Taking this dependence into account in (9.13) introduces additional terms to equation (9.9), which for an anisotropic interfacial energy becomes

$$\frac{\partial \phi}{\partial t} = M\left[W\phi(1 - \phi)\left(\phi - \frac{1}{2} - \frac{\Delta g(T)}{W}\right) + \nabla(\varepsilon^2 \nabla \phi) \right.$$
$$\left. + \left(\frac{\partial}{\partial y}\left(\varepsilon\varepsilon'\frac{\partial \phi}{\partial x}\right) - \frac{\partial}{\partial x}\left(\varepsilon\varepsilon'\frac{\partial \phi}{\partial y}\right)\right)\right] \tag{9.19}$$

where $\varepsilon' = \dfrac{\mathrm{d}\varepsilon}{\mathrm{d}\theta}$. The effect of introducing the interfacial energy can be seen by contrasting the anisotropic results of figure 9.8 with the isotropic ($\delta = 0$) results of figure 9.7. Firstly, it can be seen in figures 9.8(a) and (b) for planar growth ($K \leqslant 1$)

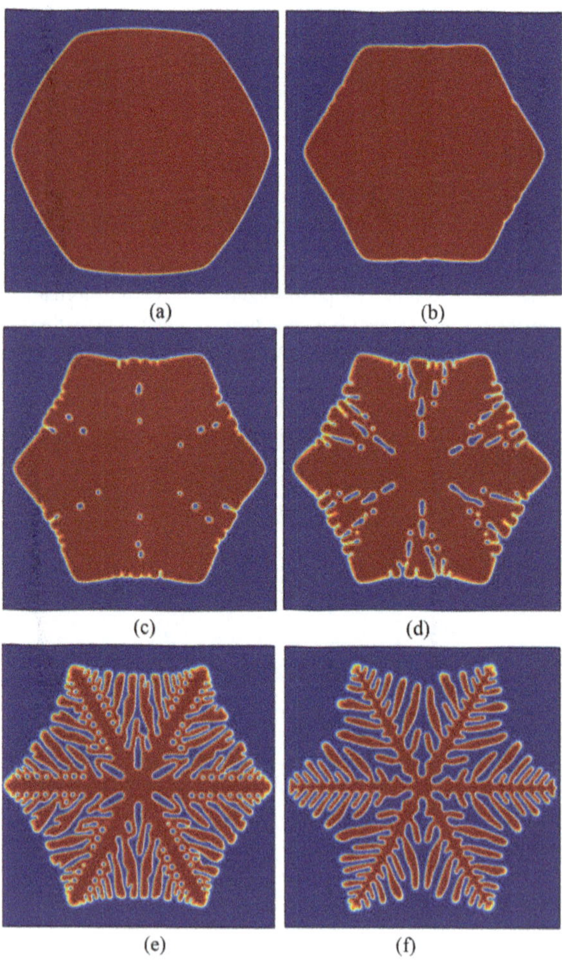

Figure 9.8. Phase field variable ϕ of a liquid (blue) to solid (red) phase transformation using equations (9.19) and (9.17) for an anisotropic interfacial energy. Growth is initiated from a small circular solid seed region in the middle. The different latent heats are: (a) $K = 0.8$, (b) $K = 1.0$, (c) $K = 1.1$, (d) $K = 1.2$, (e) $K = 1.5$, (f) $K = 2.0$. Modelling performed using COMSOL Multiphysics® [8].

that the interface now wishes to form facets in six different directions, promoting an hexagonal shape. As the latent heat measure K increases, the interface destabilises, as seen before, but now growth is strongly inclined towards one of the six preferred directions. For $K = 2$, in figure 9.8(f), the effect is most strongly seen, where a classic 'ice crystal' configuration is observed. Each of these six branches are known as **dendrites**, from the Greek meaning tree-like. Figure 9.8 shows the transition from **planar growth** to **dendritic growth** as the latent heat measure K increases.

To understand the physical origin of the interfacial instability, it is useful to observe the different temperature distributions found in the crystals for the cases of planar growth and dendritic growth, as shown in figure 9.9. For the stable case, figure 9.9(a) shows that the temperature of the solid is always below its melting temperature, i.e. $\bar{T} < 1$. The reason for this is easy to see. In the absence of heat conduction, the local increase in thermal energy of interfacial solid due to latent heat release is $\rho c_\mathrm{p} \Delta T = L_\mathrm{f}$. The critical temperature increase here is $\Delta \bar{T} = 1$, or $\Delta T = T_\mathrm{M}$, because that raises the undercooled liquid from $\bar{T} = 0$ up to the melting temperature $\bar{T} = 1$. So if $K = \dfrac{L_\mathrm{f}}{\rho c_\mathrm{p} T_\mathrm{M}} < 1$ then the latent heat release keeps the temperature of the solid below melting $(T < T_\mathrm{M})$, and if $K > 1$ then it raises the temperature above melting $(T > T_\mathrm{M})$. Equation (9.15) shows that the former case $(K < 1)$ will promote phase change, whereas the latter $(K > 1)$ will suppress it and stop growth. However, the cessation of growth for $K > 1$ can be avoided by destabilization of the interface as the best way to continue growth is to increase the interfacial area to maximise heat dissipation to the liquid, much in the same way that heat fins are placed on a radiator to increase its surface area. A small hot protuberance emanating from a planar interface will be surrounded on three sides by cool liquid. It can therefore lose heat much more quickly, hence allowing the phase transformation to continue. The dendritic morphology also means that the liquid between the dendrites heats up, suppressing the phase transformation there. This can be seen in figure 9.9(b), where the temperature is close to $\bar{T} = 1$ across the entire region encapsulated by the crystal,

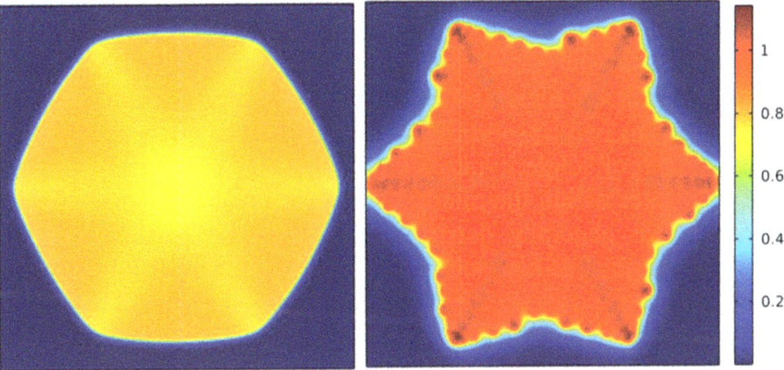

Figure 9.9. Normalised temperature distribution, $\bar{T}$ for the crystals in figure 9.8 when (a) $K = 0.8$ and (b) $K = 2.0$. The liquid is initially undercooled at $\bar{T} = 0$. The melting temperature is $\bar{T} = 1$. Modelling performed using COMSOL Multiphysics® [8].

whether liquid or solid. The interstitial liquid regions do not solidify in this simulation because no heat can leave the system, but in practice the entire system will cool over time and allow these regions to solidify as well. Here the undercooling is very large, but planar growth can be promoted by low undercooling (low solidification rate) and/or high thermal conductivity (to take the heat away from the interface more rapidly).

The model above is for a unary system. Dendritic growth is also commonly seen in binary or higher component systems. In such systems, the instability of the interface is more strongly affected by **constitutional undercooling** than thermal undercooling. In an alloy, solute is ejected from the solid as it grows, thereby enriching the solute content of the interfacial liquid. As shown in the phase diagram in figure 9.1, increasing the solute content lowers the solidification temperature of the liquid. As we have seen for precipitate growth in section 7.2.2, solute piles up in front of the interface slowing it down. However, in the case of solidification this pile-up can be avoided by destabilisation of the interface. It is unstable because a small solid protuberance grown from a planar interface will move itself out beyond the interfacial pile-up, into a region of liquid with lower solute concentration. This region will have a higher solidification temperature, hence promoting the phase transition and the further growth of the protuberance. One simple analogy is a snowplough. Think of a planar interface as a very wide flat-fronted snowplough, more like a digger, which moves forward pushing the snow (solute) in front of it. The snow gets higher and higher as it moves forward causing the plough to slow down (in this analogy the snow cannot go around the sides or over the top of the plough). Now think of a dendritic protuberance as a V-shaped snowplough where, as it pushes forwards, the snow is pushed out to the sides. In this case, the snowplough is not slowed down by the accumulation of snow in front of it, but can keep moving as efficiently as when it started.

There are simple models for predicting the onset of constitutional undercooling [6] and the conditions under which it can be avoided. However, in practice, it can only be suppressed by an extremely low solidification rate (of the order of millimetres per hour), and hence dendritic growth is fairly ubiquitous in all alloys. The consequences of this are **microsegregation** within the resulting solid. The concept of solute partitioning was introduced in section 3.2.5, and dendrite formation is largely the mechanism by which it occurs. When fully solidified, the dendritic microstructure (which solidified first) will have low solute content compared to the inter-dendritic regions (which solidified last with a higher solute content). This non-uniform distribution of solute, typically at the micron scale, leads to non-uniform mechanical properties within the microstructure, e.g. some regions will be stronger than others. This is generally detrimental to material performance. It can be ameliorated by annealing at high temperatures to allow redistribution of the solute, although this is not an economic option for rapidly produced, low cost components.

References

[1] Fredriksson H and Akerlind U 2012 *Solidification and Crystallization Processing in Metals and Alloys* (New York: Wiley)

[2] Porter D A, Easterling K E and Sherif M Y 2018 *Phase Transformations In Metals And Alloys* 3rd edn (Boca Raton, FL: CRC Press)

[3] Takaki T 2014 Phase-field modeling and simulations of dendrite growth *ISIJ Int.* **54** 437–44

[4] Tourret D, Liu H and Lorca L J 2022 Phase-field modeling of microstructure evolution: Recent applications, perspectives and challenges *Prog. Mater Sci.* **123** 100810

[5] Kobayashi R 1993 Modeling and numerical simulations of dendritic crystal growth *Phys.* D **63** 410–23

[6] DoITPoMs solidification page. https://doitpoms.ac.uk/tlplib/solidification_alloys/index.php

[7] Pandat: software suite for thermodynamic calculation and kinetic simulation of multi-component alloys. CompuTherm LLC, Madison, Wisconsin, USA www.computherm.com [accessed 1 October 2021]

[8] COMSOL Multiphysics® v. 5.6. COMSOL AB, Stockholm, Sweden www.comsol.com [accessed 1 October 2021]

Chapter 10

Exercises

1. Create a fictitious phase diagram with five phases. Start with the basic eutectic phase diagram produced by 'twophase.tdb' at the start of chapter 3 with $L_S = 20$ kJ mol^{-1} and $L_L = 0$ J mol^{-1}. Add code to the TDB file to create three more stoichiometric phases, AB, AB$_2$ and A$_3$B with $g_{AB2} = -1.2$ kJ mol^{-1}, $g_{AB} = -1$ kJ mol^{-1} and $g_{A3B} = -1.5$ kJ mol^{-1}. Label the phase diagram and explore the origin of the phase boundaries by plotting relevant isothermal energy profiles.

2. The Gibbs free energy profiles of three phases are given as a function of composition at a selection of temperatures in figure E.1. Identify the single and two-phase regions on each diagram and use these to sketch the binary phase diagram for this system.

3. The Gibbs free energy profiles of four phases are given as a function of composition at a selection of temperatures in figure E.2. Identify the single and two-phase regions on each diagram and use these to sketch the binary phase diagram for this system. In this example there are two stoichiometric phases.

4. Download the Fe–C TDB file from the NIMS database (see figure 1.5) then, using PANDAT, calculate the energetic driving force for nucleation $\Delta g_{L \to S}$ in a liquid-solid region of the iron-carbon system, using the graphical methodology of figure 5.2, for a temperature of $T = 1500$ °C and a nominal composition of 1 mol% C. Check your answer against the numerical result returned by PANDAT. (Hint: Firstly determine the solid phase that is

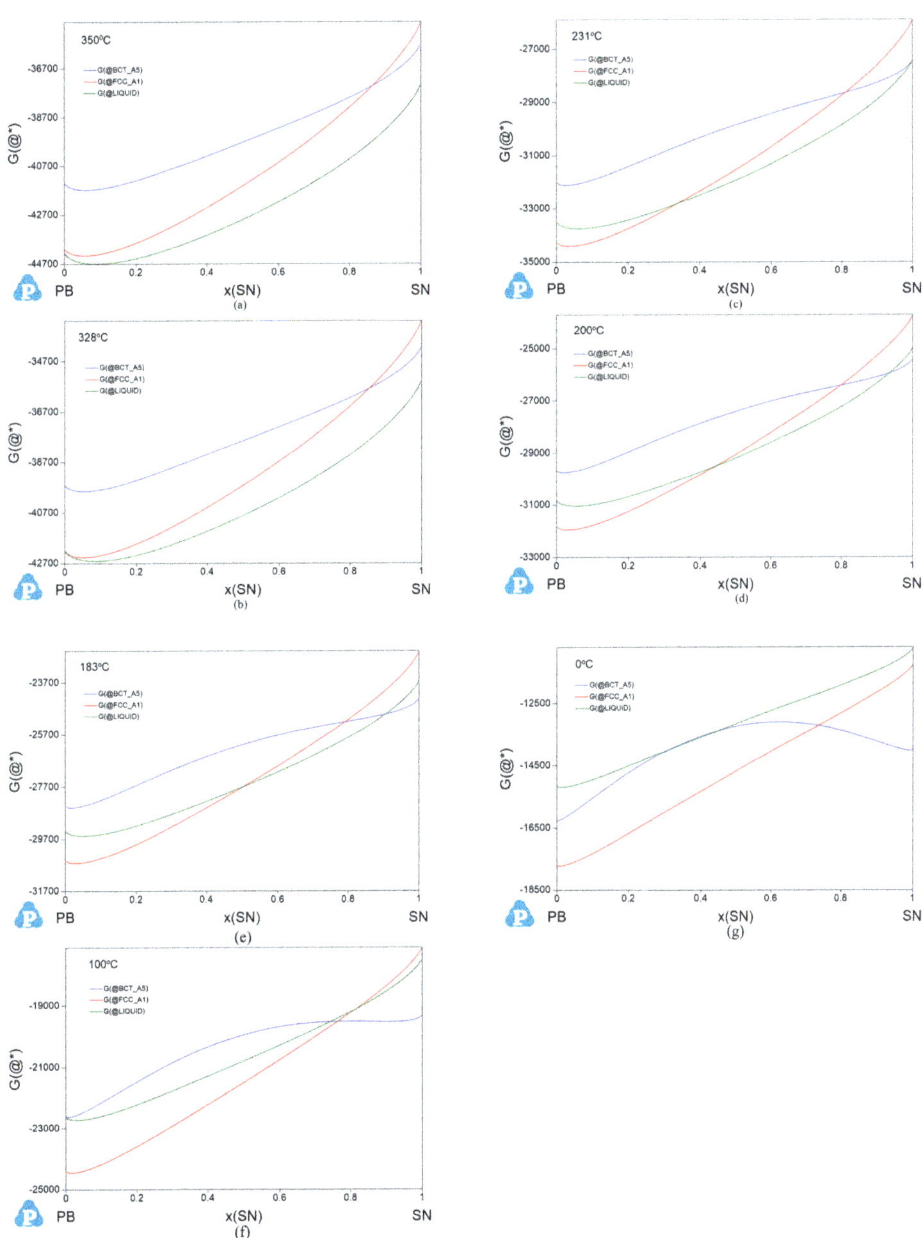

Figure E.1. Gibbs free energy profiles for a three phase, binary system as a function of composition at seven selected temperatures. Created using PANDAT [1].

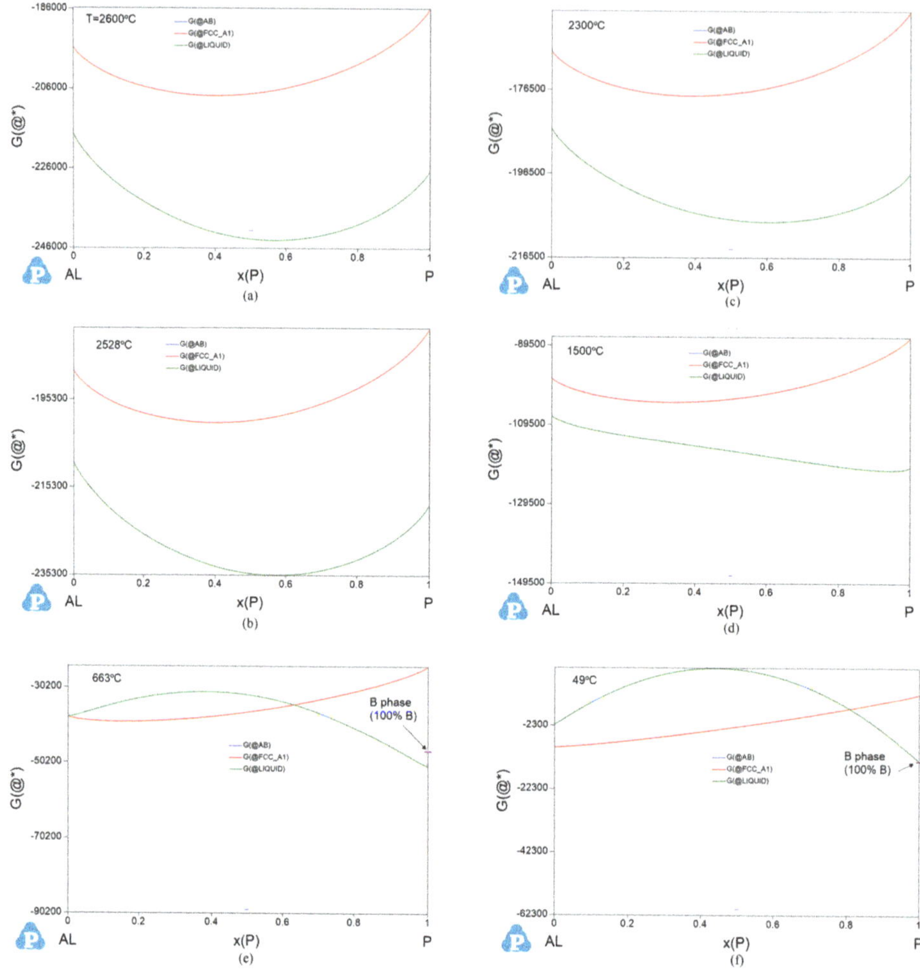

Figure E.2. Gibbs free energy profiles for a four phase, binary system as a function of composition at six selected temperatures. There are two stoichiometric phases. Created using PANDAT [1].

expected to appear. You will need to zoom into a small region of the phase diagram to see this. Only plot the energy for the relevant phases over the compositional range of interest to avoid confusion. Use the line drawing tool to sketch the construction lines on the relevant Gibbs free energy LINE plot.)

5. Figure E.3 shows the Al–Mg–Zn ternary liquidus lines. Determine the two liquidus compositions for a 20 mol% Zn alloy by drawing construction lines on the diagram below.

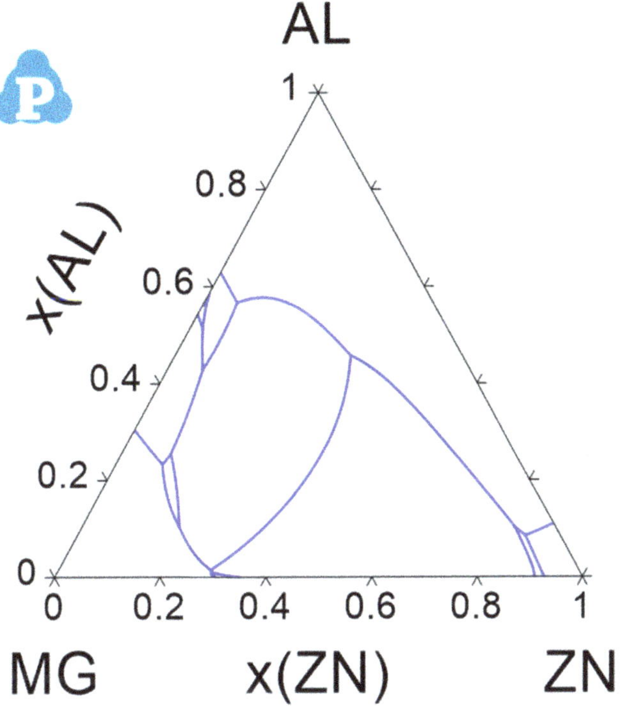

Figure E.3. Liquidus lines for the Al–Mg–Zn ternary system. Created using PANDAT [1].

References

[1] Pandat: software suite for thermodynamic calculation and kinetic simulation of multi-component alloys. CompuTherm LLC, Madison, Wisconsin, USA www.computherm.com [accessed 1 October 2021]

Chapter 11

Solutions

1. The TDB code for the 'fivephase.tdb' phase diagram and the associated labelled phase diagram in figure S.1.

```
ELEMENT A SOLID 1.0 0.0 0.0 !
 ELEMENT B SOLID 1.0 0.0 0.0 !

 PHASE SOLID % 1 1 !
 CONSTITUENT SOLID : A B : !
 PARAMETER G(SOLID,A) 298.15 0.0; 3000.0 N !
 PARAMETER G(SOLID,B) 298.15 0.0; 3000.0 N !
 PARAMETER L(SOLID,A,B;0) 298.15 20000.0; 3000.0 N !

 PHASE LIQUID:L % 1 1 !
 CONSTITUENT LIQUID:L : A B : !
 PARAMETER G(LIQUID,A) 298.15 20.0*(800.0+273.15-T);
3000.0 N !
 PARAMETER G(LIQUID,B) 298.15 20.0*(1000.0+273.15-T);
3000.0 N !
 PARAMETER L(LIQUID,A,B;0) 298.15 0.0; 3000.0 N !

PHASE AB % 2 1 1 !
 CONSTITUENT AB : A : B : !
 PARAMETER G(AB,A:B) 298.15 -1000.0; 3000.0 N !
```

doi:10.1088/978-0-7503-3147-0ch11

```
PHASE AB2 % 2 1 2 !
 CONSTITUENT AB2 : A : B : !
 PARAMETER G(AB2,A:B) 298.15 -1200.0; 3000.0 N !

PHASE A3B % 2 3 1 !
 CONSTITUENT A3B : A : B : !
 PARAMETER G(A3B,A:B) 298.15 -1500.0; 3000.0 N !
```

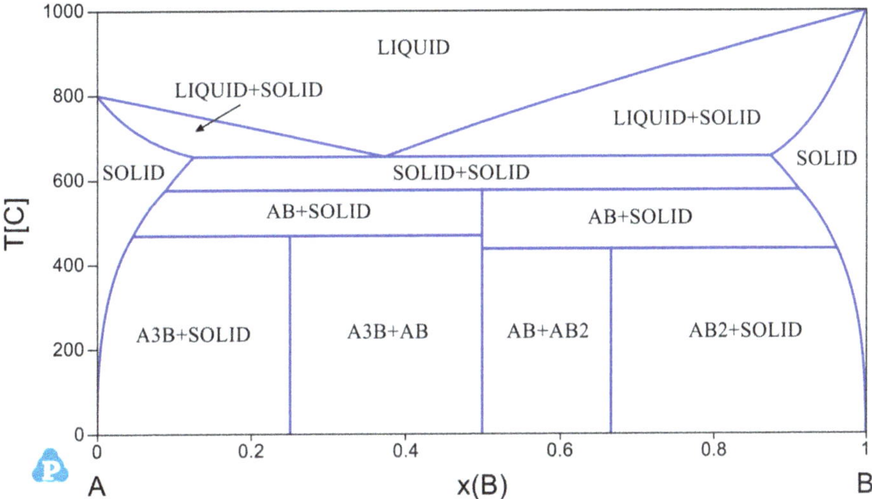

Figure S.1. Solution to the Q1 binary phase diagram. Created using PANDAT [1].

2. The binary system is Pb–Sn. Download the TDB file from the NIMS database (see figure 1.5), generate and label the phase diagram in PANDAT and check the result against your own. Figure S.2 adds markings to the plots of figure E.1 to define the single and two phase regions at each temperature. These boundary points at each temperature are then transposed on to a new graph, with the regions clearly labelled on each isotherm. The points are then connected to sketch the phase boundaries, as shown in the final plot.

3. The binary system is Al–P. Download the TDB file from the NIMS database (see figure 1.5), generate and label the phase diagram in PANDAT and check the result against your own. Figure S.3 adds markings to the plots of figure E.2 to define the single and two phase regions at each temperature. These boundary points at each temperature are then transposed on to a new graph, with the regions clearly labelled on each isotherm. The points are then connected to sketch the phase boundaries, as shown in the final plot. The AB phase is AlP and the B phase is white phosphorus.

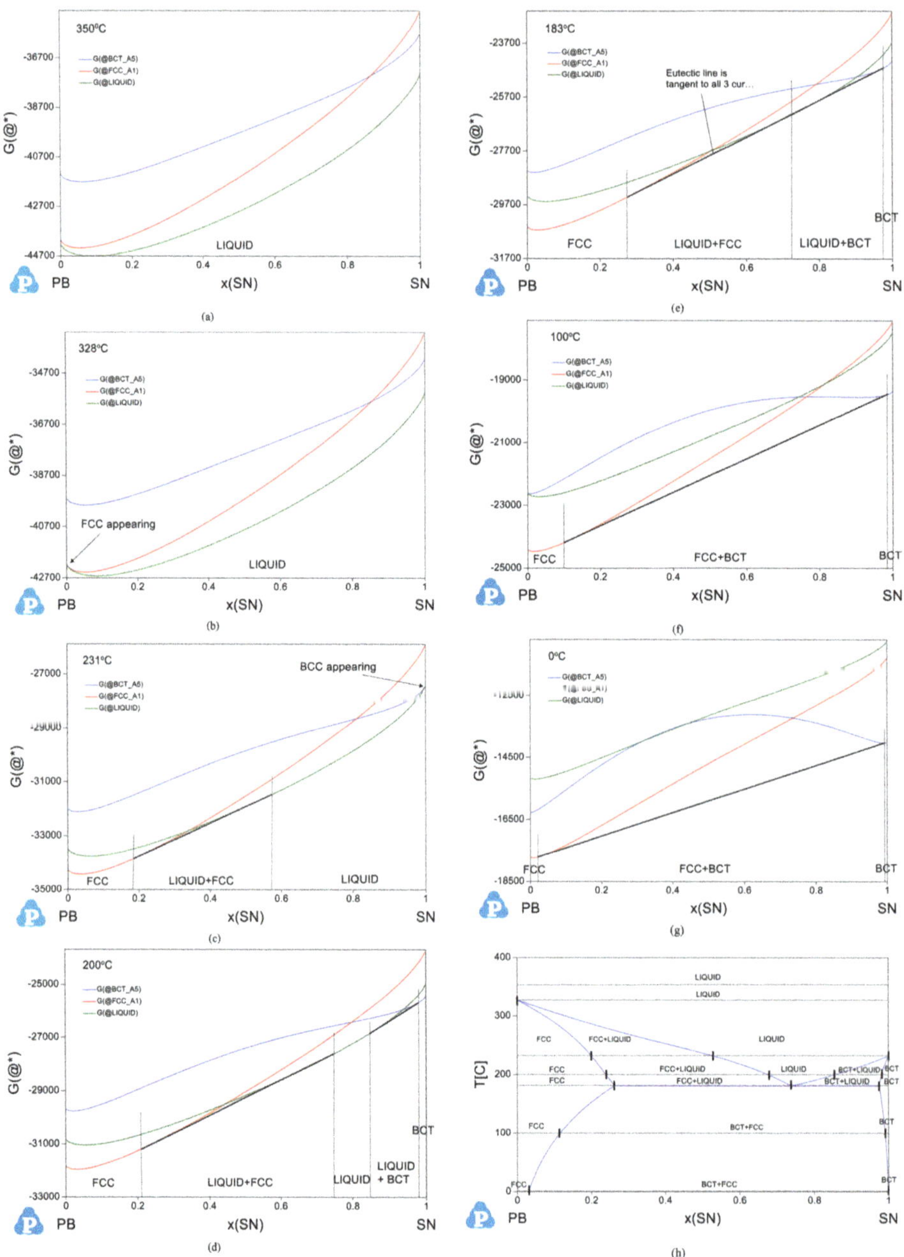

Figure S.2. Solution to the Q2 binary phase diagram, showing single and two phase regions on Gibbs free energy plots at seven different temperatures, and finally the construction of the phase diagram. Created using PANDAT [1].

Figure S.3. Solution to the Q3 binary phase diagram, showing single and two phase regions on Gibbs free energy plots at six different temperatures, and finally the construction of the phase diagram. Created using PANDAT [1].

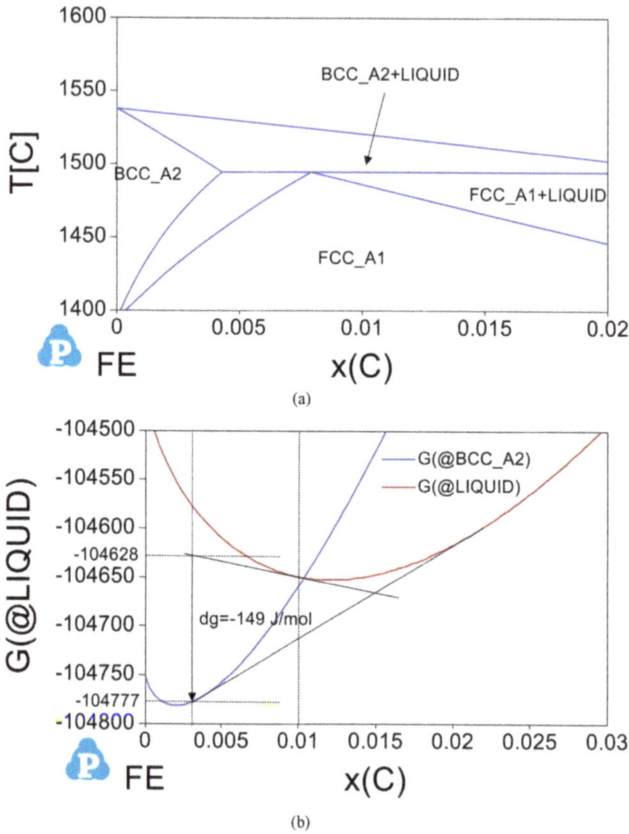

Figure S.4. (a) The area of the Fe–C phase diagram of interest shows that the BCC and LIQUID phases are predicted at 1 mol% C at 1500 °C. (b) The Gibbs free energies of the two relevant phases are plotted over the relevant compositional range. The construction lines are drawn in accordance with the method shown in figure 5.2. Created using PANDAT [1].

4. A small area of the phase diagram is constructed to determine the relevant solid phase. This is shown in figure S.4(a). The Gibbs free energies of the BCC_A2 and LIQUID phases are then plotted on their own in figure S.4(b), to avoid the confusion of the other phases. The construction lines are drawn, in accordance with the method shown in figure S.5(a). Firstly, draw a line that is tangential to both the liquid and solid curves. Then draw a line vertically at 0.003, where this line intersects the solid curve. Then draw a line vertically at the nominal composition of 0.01. Then, where this line intersects the liquid curve, draw a tangent to the liquid curve. The change in energy between where this line intersects the 0.003 vertical, and where it touches the solid curve is the change in energy. The estimated

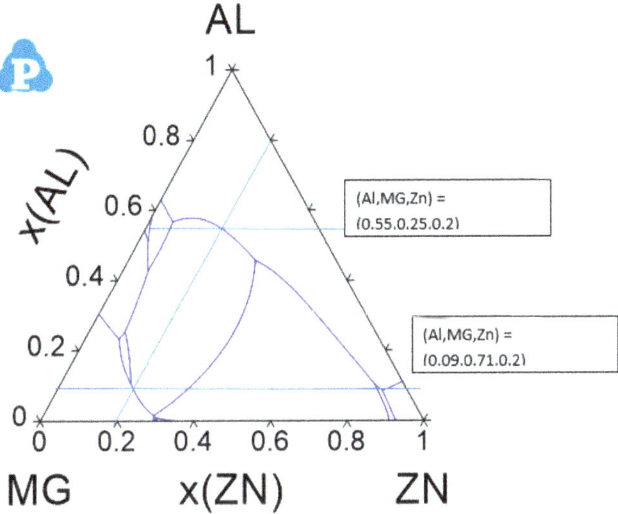

Figure S.5. Eutectic compositions for 20% mol Zn. Created using PANDAT [1].

driving force for the liquid-to-solid phase transformation from the diagram is 149 J mol^{-1}. The numerical value of DF, calculated using PANDAT using the process described in section 5.2, is 157.148 J mol^{-1}. The error in the estimated value arises from visual approximation of the tangent lines.

5. Firstly, the 0.2 Zn construction line is drawn, as shown in figure S.5. The allowable range of Al–Mg compositions lies along this line. It crosses the liquidus lines twice. These are the eutectic points, at the compositions identified.

Reference

[1] Pandat: software suite for thermodynamic calculation and kinetic simulation of multi-component alloys. CompuTherm LLC, Madison, Wisconsin, USA www.computherm.com [accessed 1 October 2021]